walkermaths 1.11

BIVARIATE DATA

NCEA Level 1 Internal

Walker Maths 1.11 Bivariate Data
1st Edition
Charlotte Walker
Victoria Walker

Editor: Eva Chan
Designer: Cheryl Smith, Macarn Design
Production controller: Siew Han Ong

Acknowledgements
Cover photo courtesy of Shutterstock.

We wish to thank the Boards of Trustees of Darfield and Riccarton High Schools for allowing us to use materials and ideas developed while teaching. Our thanks also go to all past and present colleagues who have generously shared their expertise and ideas.

For product information and technology assistance,
in Australia call **1300 790 853**;
in New Zealand call **0800 449 725**

For permission to use material from this text or product, please email **aust.permissions@cengage.com**

National Library of New Zealand Cataloguing-in-Publication Data
A catalogue record for this book is available from the National Library of New Zealand.

978 0 17 037163 6

Cengage Learning Australia
Level 7, 80 Dorcas Street
South Melbourne, Victoria Australia 3205

Cengage Learning New Zealand
Unit 4B Rosedale Office Park
331 Rosedale Road, Albany, North Shore 0632, NZ

For learning solutions, visit **cengage.co.nz**

Printed in China by 1010 Printing International Limited
11 12 13 14 25 24 23 22

CONTENTS

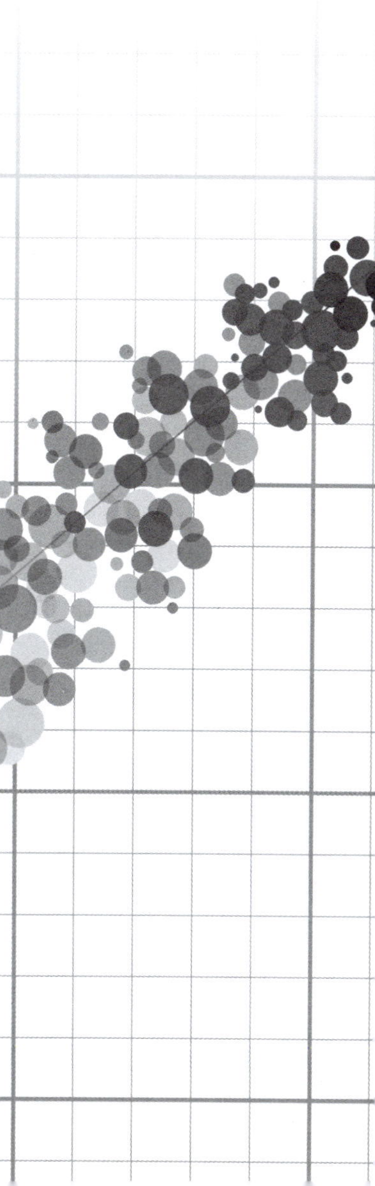

ISBN: 9780170371636

Glossary

Make your own glossary of key terms:

Term	Definition	Picture/Example
Bivariate		
Sample		
Population		
Trendline		
Group		
Unusual point		
Inference		
Prediction		

ISBN: 9780170371636

Introduction

This standard will require you to investigate relationships (if any) between sets of data. More specifically you will be required to:

- plan and conduct an investigation involving bivariate numerical data
- work with a given relationship question
- determine appropriate variables and measures
- manage sources of variation
- gather data
- select and use appropriate display(s)
- communicate relationship(s) in the data in a conclusion.

What is bivariate data?

- This is where you have **two** pieces of **numeric** (number) data about every individual in a sample or population.
- You analyse **both** at once to see if there is a relationship between them.
- These are plotted on a **scatter** graph.

Statistical enquiry cycle

Your report should connect back to all aspects of the statistical enquiry cycle:

ISBN: 9780170371636

Variables

There are three types of variables:

Continuous variables These are **measured** values which usually include **fractions or decimals**.
Examples: height, weight, distance, time.

Discrete variables These are **counted** values which are usually **whole** numbers.
Examples: number of pets, shoe size (although this is unusual because these can be in halves).

Descriptive variables These are **categorical** variables which are usually **words**.
Example: eye colour, type of pet.

In theory we should analyse only **continuous** bivariate data. However, in practice, provided there is a large enough range, we sometimes analyse bivariate discrete data, e.g. marks out of 100.

The following information was collected from each member of a Year 11 class. Complete the table by stating what type of variable each is, and whether or not (✓ or ✗) each would be a suitable variable to use as part of a bivariate study.

Variable	Type of variable	Suitable for a bivariate study?
Favourite colour	Descriptive	X
Foot length		
Number of pens in pencil case		
Head circumference		
Shoe size		
Number of pets		

 ISBN: 9780170371636

Variable	Type of variable	Suitable for a bivariate study?
Number of text messages on phone		
Favourite ice-cream flavour		
Height		
Area of bedroom floor		
Distance travelled to school		
Eye colour		
Arm span		
Make of calculator		
Age		
Time taken to eat a dry Weet-Bix		
Amount spent in the canteen last week		
Weight of school bag		

From this list, select some pairs of variables that would be suitable for a bivariate investigation:

Height	and	Arm span
	and	
	and	
	and	
	and	
	and	

ISBN: 9780170371636

Display of data

General guidelines for drawing graphs

- Your graph should be at least **half a page**.
- It should have a **title**.
- Your axes should be **ruled**.
- Scales must be **labelled** with what is being measured and the units used.
- Scales must be **regular**.

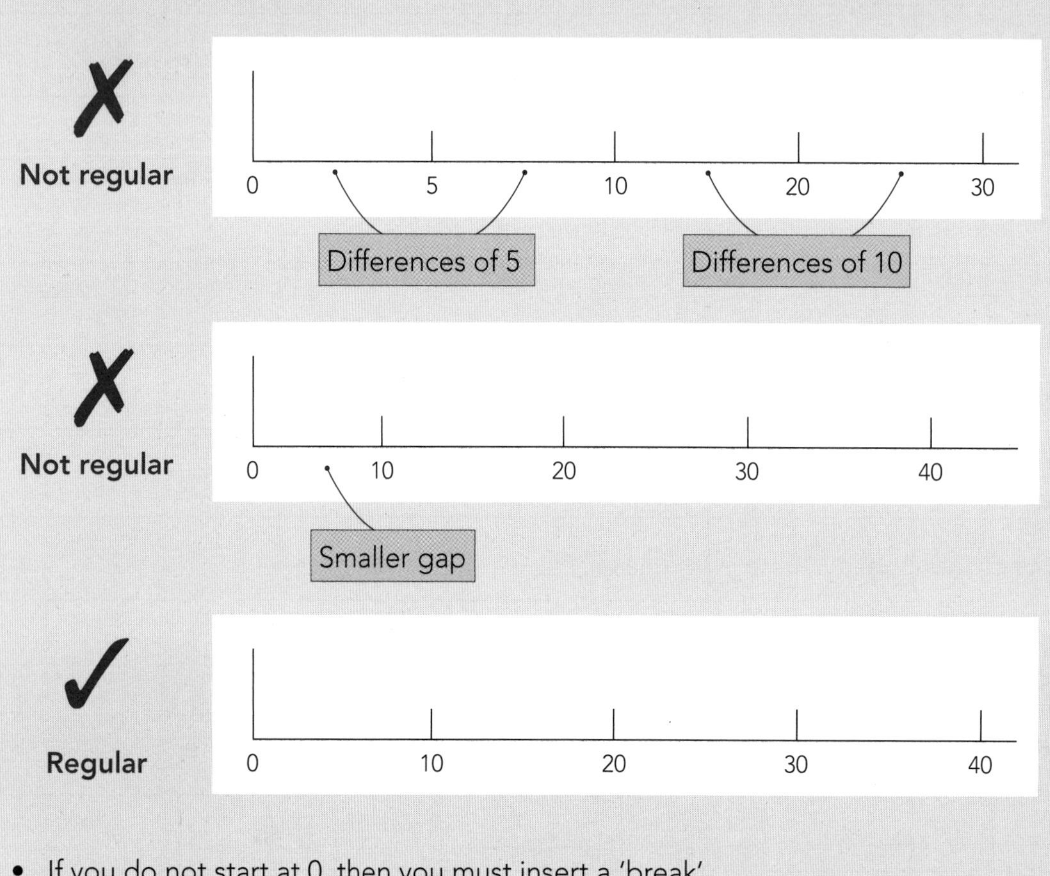

- If you do not start at 0, then you must insert a 'break'.

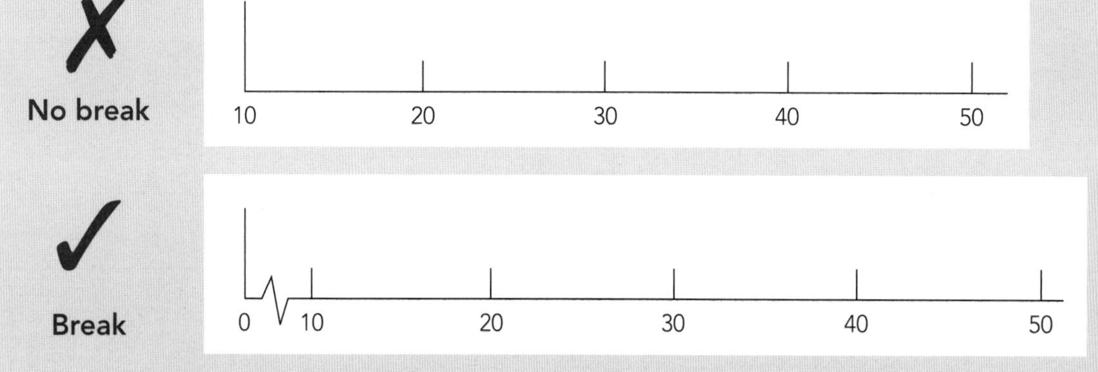

ISBN: 9780170371636

Creating scatter plots

- Scatter plots should be approximately square.

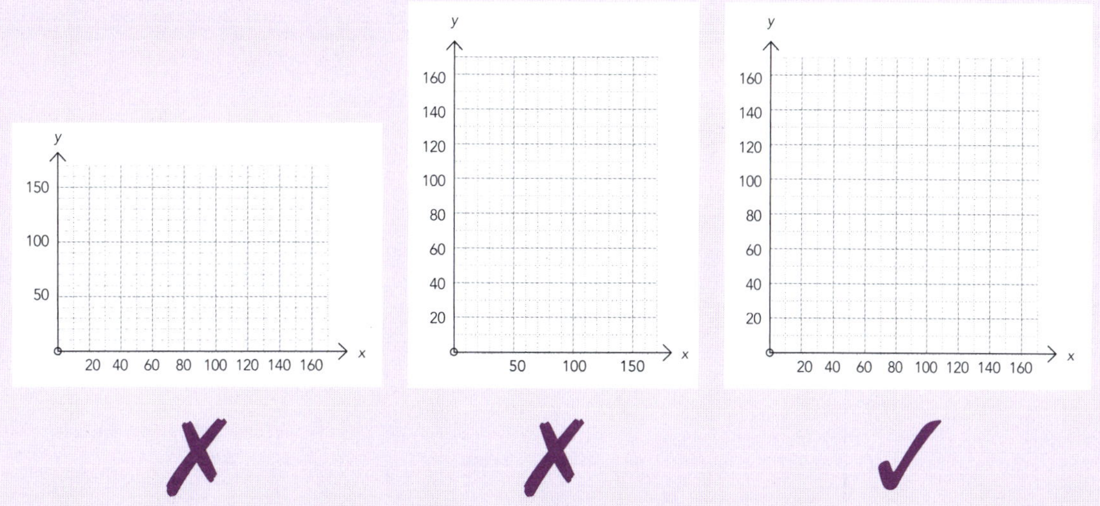

- You do not need to use the same scale on both axes.
- Each point should be marked with a clear dot or a cross.
- Make sure that you keep the pairs together.
- If you are using *x*- and *y*-axes, remember that co-ordinates are in alphabetical order: (*x*, *y*).
- If you are not using *x*- and *y*-axes, then the order is (horizontal, vertical).

Example: Plot the following points on the graph: (150, 30), (60, 120), (0, 150), (80, 0).

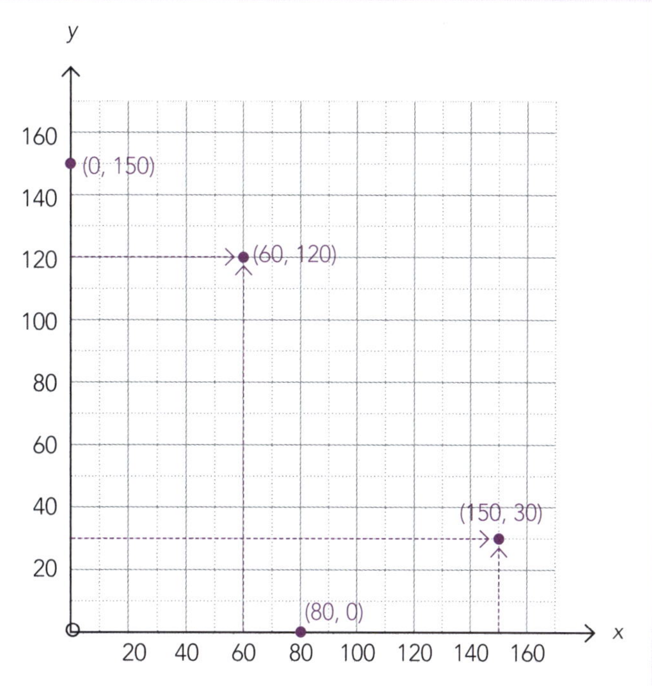

ISBN: 9780170371636

Plot the following points on the graphs below.

1 **(50, 100) (20, 160) (110, 40) (20, 0) (0, 90) (30, 40) (160, 120) (90, 140)**

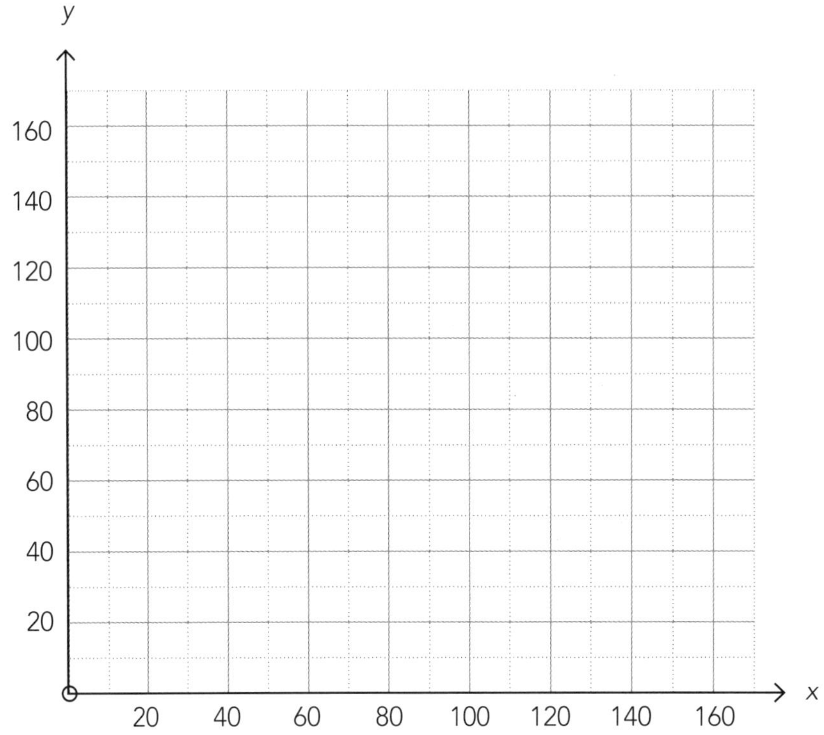

2 **(10, 30) (40, 120) (46, 90) (50, 40) (30, 0) (0, 145) (35, 190) (20, 50) (18, 170)**

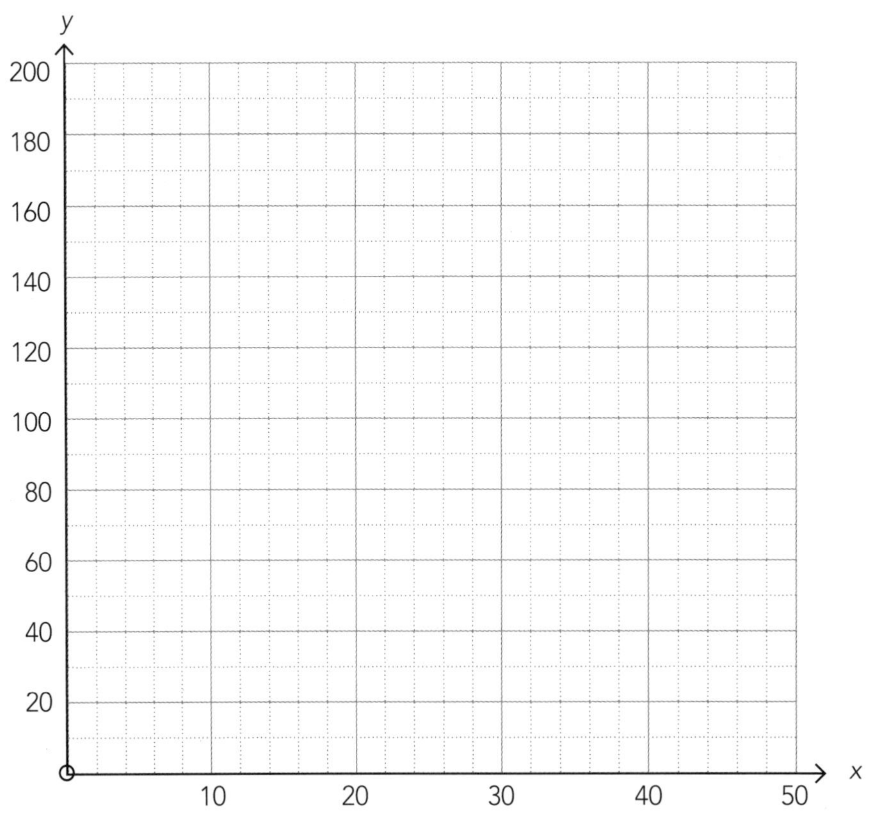

 ISBN: 9780170371636

3 (1.2, 0.3) (0, 1.9) (0.5, 3.6) (1.7, 0) (1.9, 2.3) (0.4, 0.6) (0.8, 1.6) (1.5, 3.3)

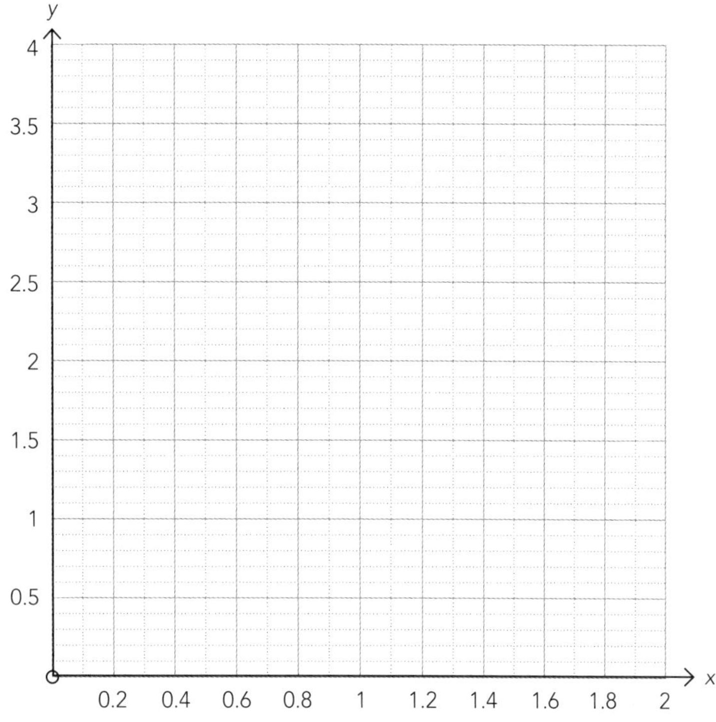

4 Use a ruler to join the points in the listed order. They should form a familiar symbol. (Hint: look at the front cover of this book).

(7, 12) (7, 6) (2, 11) (2, 17) (7, 12) (12, 17) (17, 12) (22, 17) (22, 11) (17, 6) (17, 12) stop.
(4, 9) (2, 7) (2, 1) (7, 6) (12, 1) (17, 6) (22, 1) (22, 7) (20, 9) stop.
(12, 1) (12, 7) (10, 9) (12, 11) (14, 9) (12, 7) stop. (7, 6) (10, 9) stop.
(17, 6) (14, 9) stop. (12, 11) (12, 17) stop.

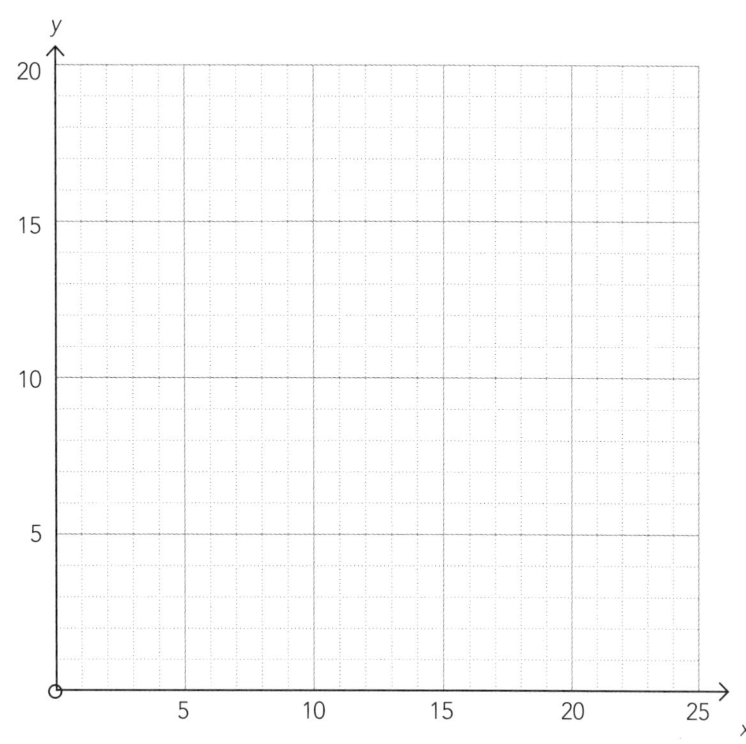

ISBN: 9780170371636

Reading scatter plots

Reading a graph value

This is done in the same way as for any other graph: (*x*-co-ordinate, *y*-co-ordinate)
or: (horizontal, vertical).

Examples:

Point **P**: (6.5, 4)

Point **Q**: (9, 9.5)

For each graph, write down the co-ordinates of the purple points and answer the questions:

1 The graph shows the relationship between students' arm span (cm) and height (cm).

a ____________________

b ____________________

c ____________________

d Give the co-ordinates of the student whose arm span and height are the same.

____

e Give the co-ordinates of the student who has very long arms compared with their height.

____

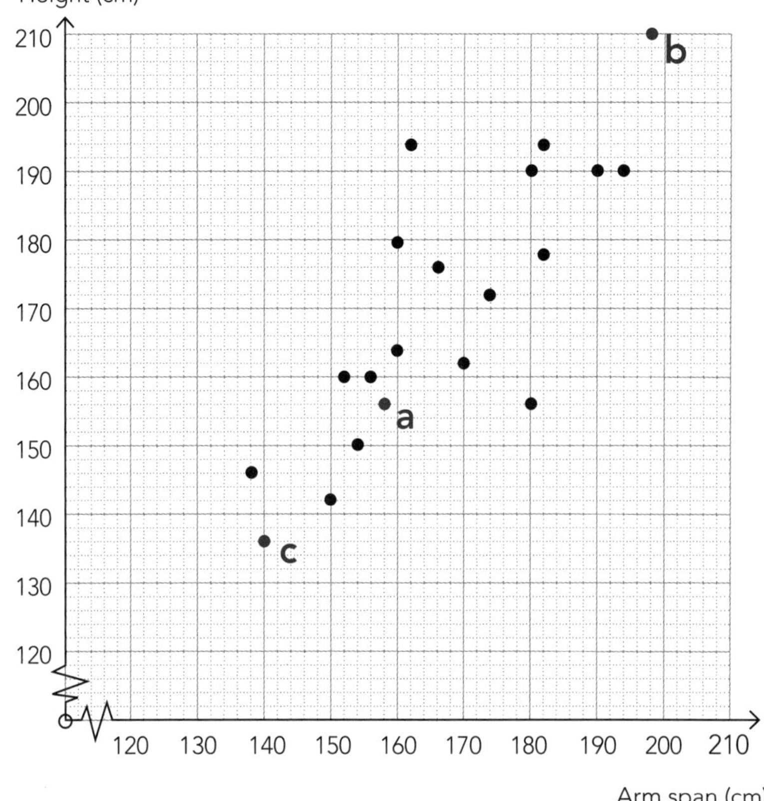

 ISBN: 9780170371636

2 The graph shows the relationship between students' height (cm) and their accuracy in a basketball goal-shooting test (%).

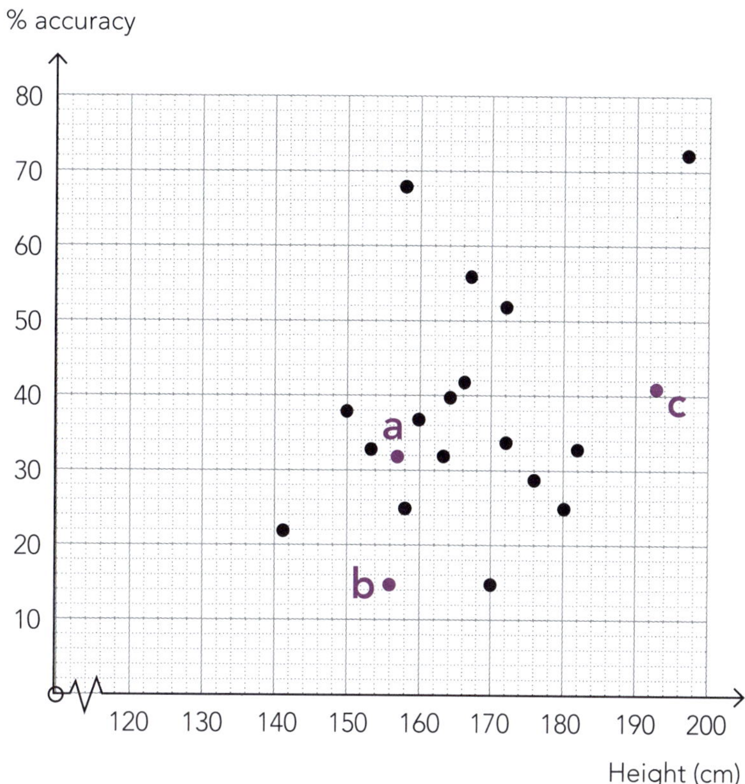

a ______________________

b ______________________

c ______________________

d How tall was the most accurate shooter?

____

e What was the accuracy of the student who was 182 cm tall?

____

3 The graph shows the relationship between the number of hours per week spent playing video games and the number of credits earned at Level 1 by a group of students.

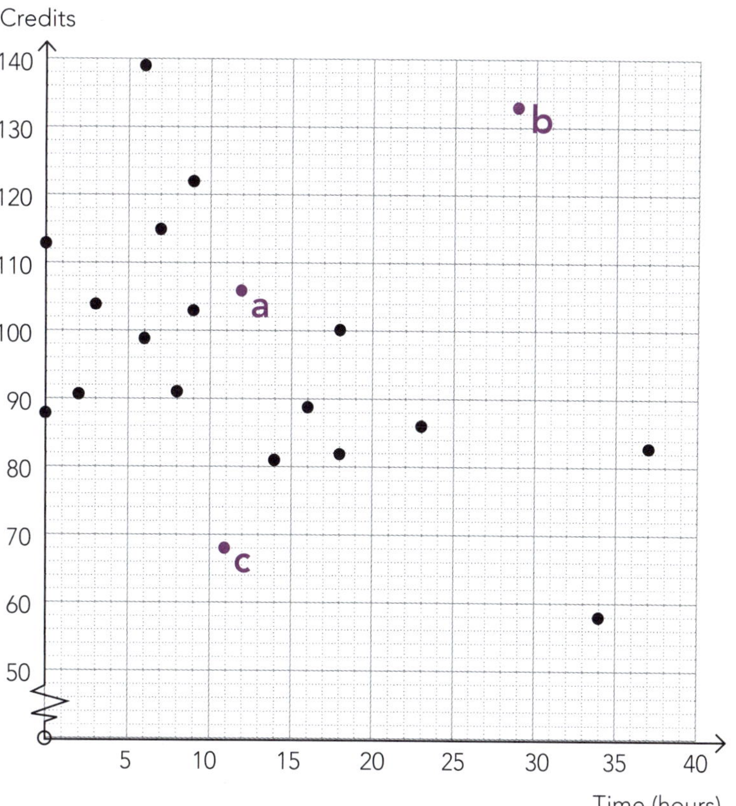

a ______________________

b ______________________

c ______________________

d How many credits were earned by the student who spent 14 hours per week playing video games?

____

e One student got 103 credits. For how many hours per week did this student play video games?

____

ISBN: 9780170371636

Drawing a line of best fit

- It is useful to draw a line of best fit before discussing a bivariate relationship.
- This line should:
  - be **ruled**
  - have about the same number of points above and below it
  - pass through the middle of the data at both ends.

Examples:

Good line of best fit:

Poor lines of best fit:

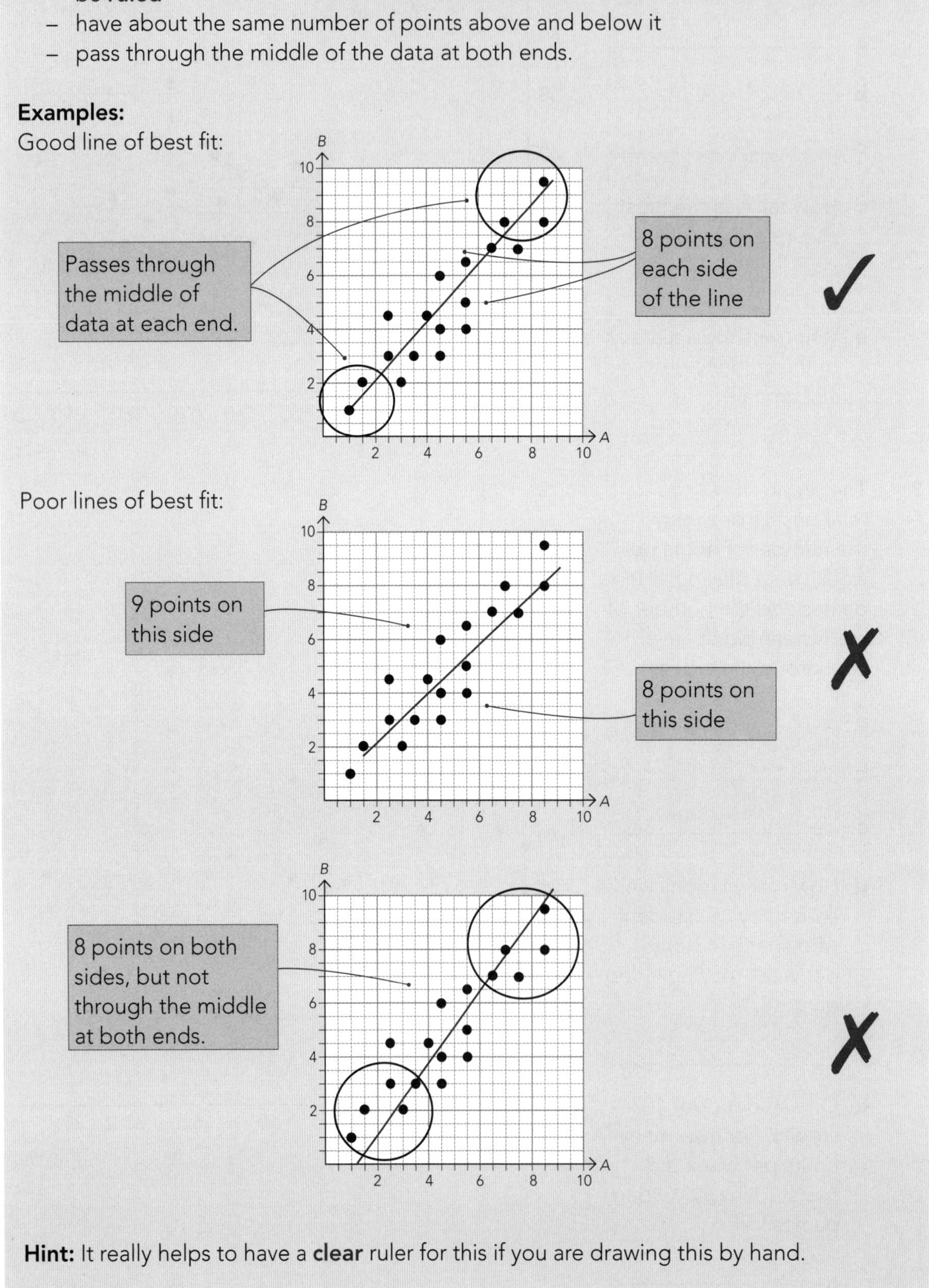

Hint: It really helps to have a **clear** ruler for this if you are drawing this by hand.

ISBN: 9780170371636

Draw lines of best fit on the following graphs.

1

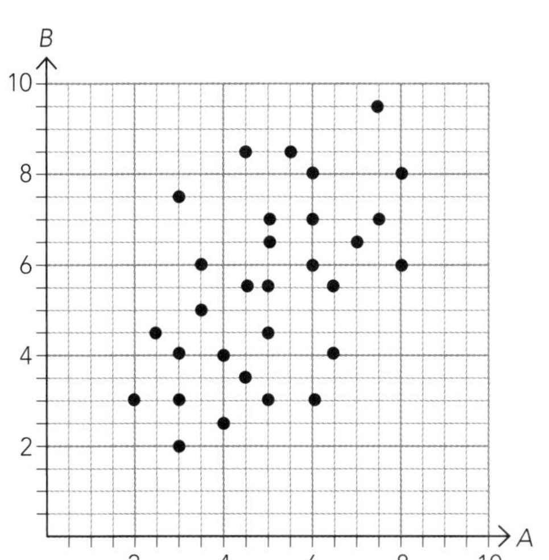

2

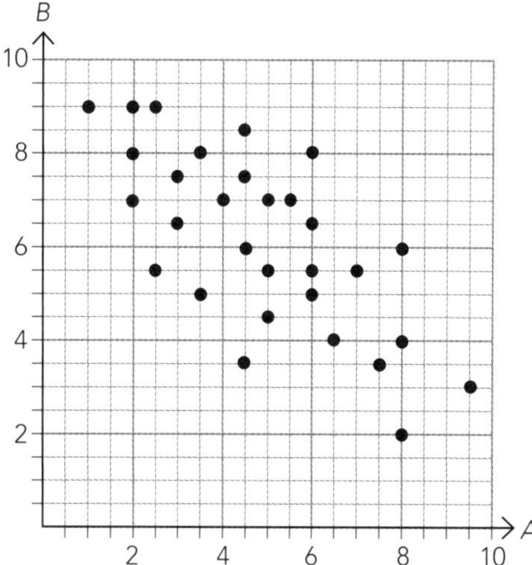

3

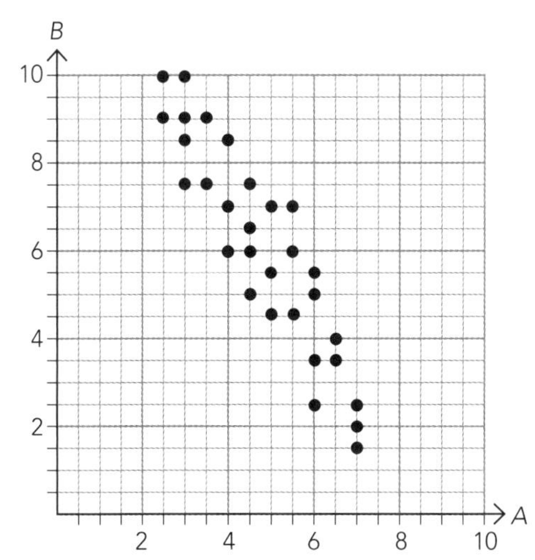

4

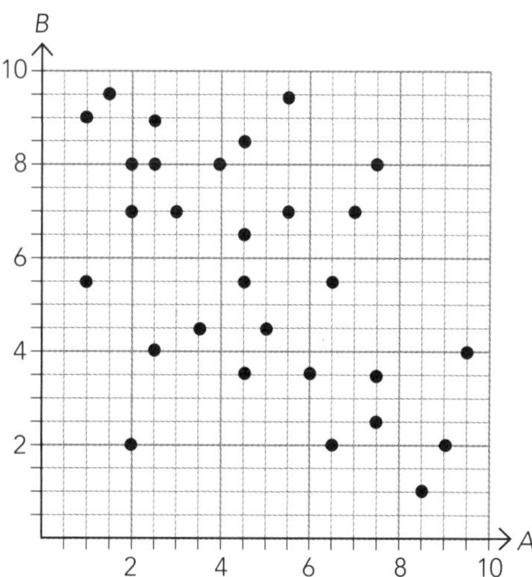

5

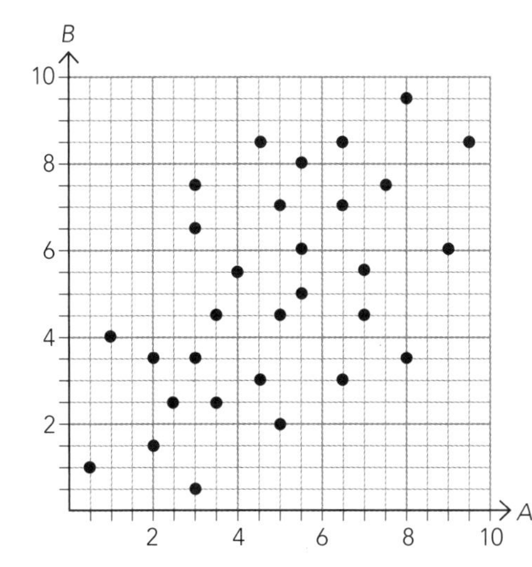

6

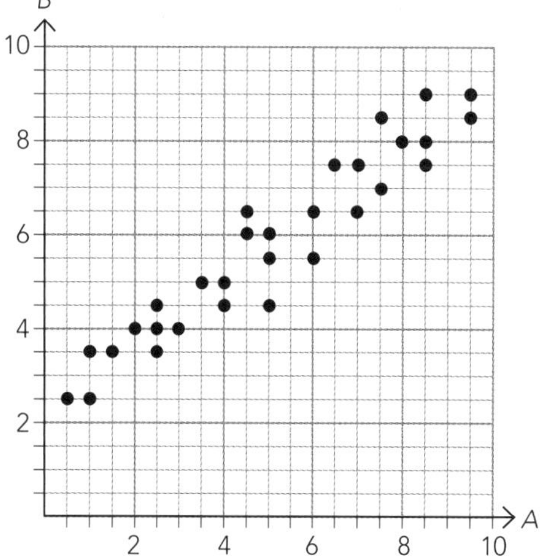

ISBN: 9780170371636

Estimating values from a line of best fit

- Once you have drawn a line of best fit, it is often useful to be able to read values from it.
- Values along the line of best fit give you 'expected' values or predictions.

Example 1:

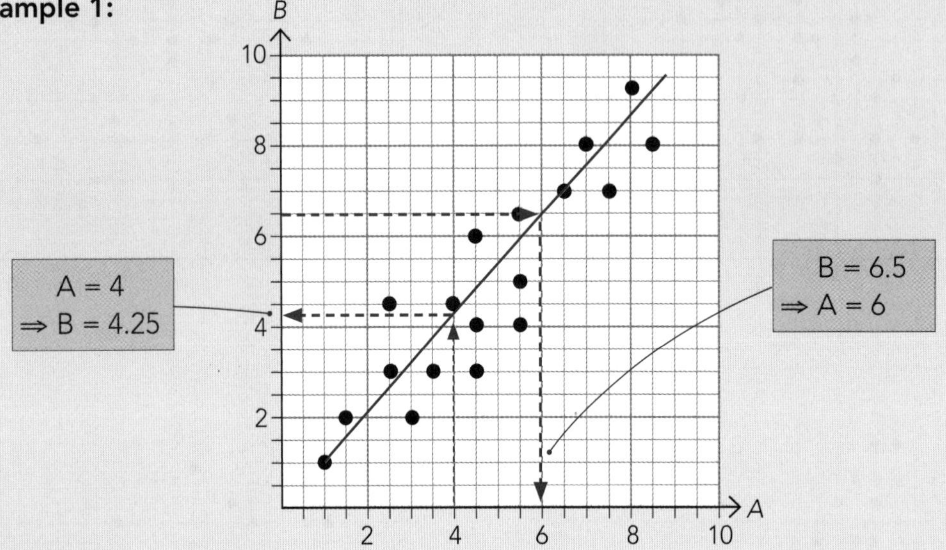

Example 2: The graph shows the relationship between students' arm spans (cm) and heights (cm).

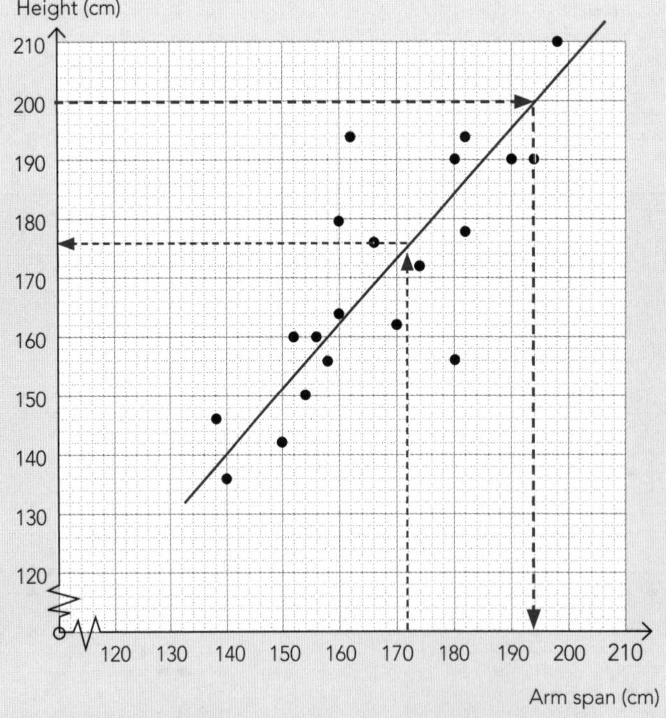

a Estimate the height of a student who has an arm span of 172 cm.
Estimated height = 176 cm

b Estimate the arm span of a student whose height is 200 cm.
Estimated arm span = 194 cm

ISBN: 9780170371636

Use the following graphs to make estimations.

1 The graph shows the relationship between head circumference and best long-jump distance for some Year 11 students.

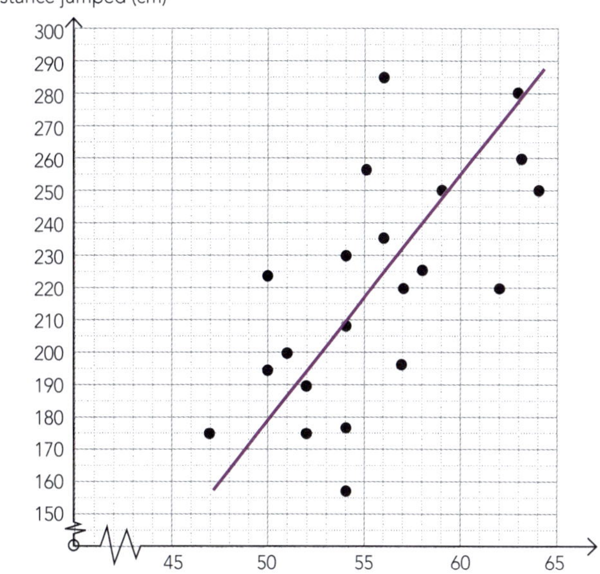

a A Year 11 student has a head circumference of 50 cm. Estimate their best long-jump distance. ____________

b A Year 11 student had a best long-jump distance of 255 cm.

Estimate their head circumference. ____________

c A Year 11 student has a head circumference of 64 cm. Estimate their best long-jump distance. ____________

2 The graph shows the relationship between students' heights (cm) and their accuracy in a basketball goal-shooting test (%).

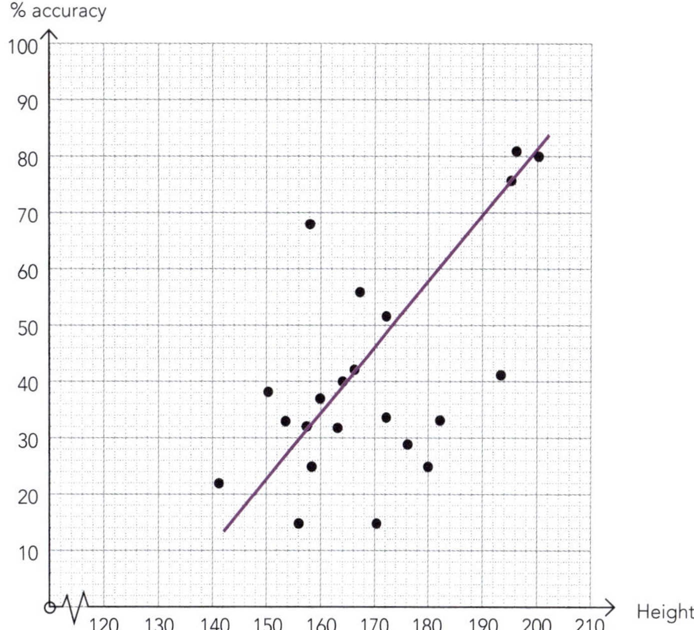

a A student's height is 170 cm. Estimate their percentage accuracy at goal shooting.

____

b A student has a goal-shooting average of 72%.

Estimate their height. ____________

c A student has a goal shooting average of 18%. Estimate their height.

____

ISBN: 9780170371636

3 The graph shows the relationship between average body weight and maximum lifespan for different breeds of dogs.

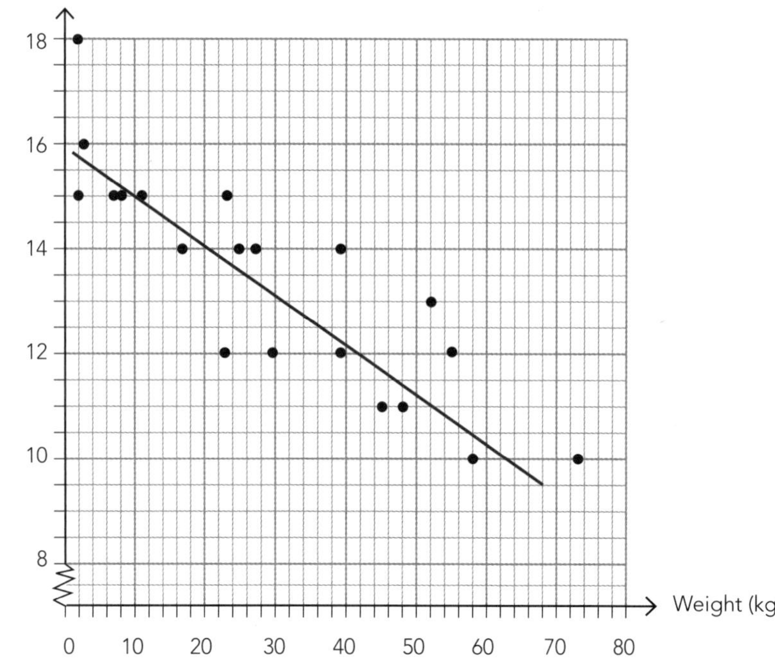

a Estimate the lifespan of a dog that weighs 26 kg.

____

b Estimate the weight of a dog that has a lifespan of 15 years.

____

c Estimate the lifespan of a dog that weighs 68 kg. ____________________

4 The graph shows the body length of some species of birds (mm) and the incubation period for their eggs (the number of days between laying the egg and hatching).

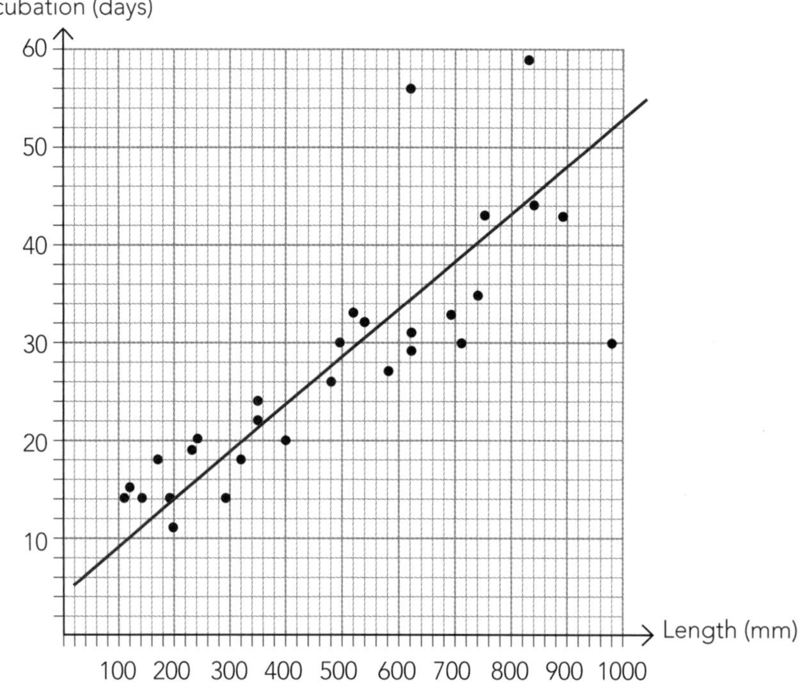

a Estimate the incubation period for a bird that is 440 mm long. ____________________

b Estimate the length of a bird that has an incubation period of 48 days. ____________________

c Which of these estimates do you think is likely to be more accurate? ____________________

Why? ____

ISBN: 9780170371636

Planning an investigation

Deciding what variables to use and why

- You will be given a question to investigate.
- You will need to:
  - state **what** variables you will use
  - state what **units** you will use
  - explain **why** you have chosen each variable.

Example: Question: What is the relationship between height and arm span of Year 11 students at Paradise High School?

For the heights:
What: *For each selected student, I will measure the height of the top of their head above the floor while they are standing against the wall.*
Units: *Centimetres.*
Why: *This will give me a measure of their overall height.*

For the arm spans:
What: *I will also measure the distance from the tip of the longest finger on their left hand to the tip of the longest finger on their right hand while they stretch their arms out horizontally and lean against the board.*
Units: *This will also be measured in centimetres.*
Why: *This will give me a measure of arm span.*

Describing how you will measure the variables and manage variation

- You will need to describe in detail **how** you will measure your chosen variables.
- In your description, you must include how you will **manage variation**.

Example: Question: What is the relationship between height and arm span of Year 11 students at Paradise High School?

For the heights:
How*: I will fasten a tape to the wall with zero on the floor. Each student will stand with their back to the wall and I will measure the distance to top of their heads.*
Managing variation:

- *I will make sure each student has their shoes off because shoes have different sole thicknesses, so to leave them on would introduce variation.*
- *I will measure the top of each student's head using a book to ensure that the height is measured at right angles to the wall.*
- *I will ask each person to stand as straight and tall as possible while keeping their feet flat on the floor.*

ISBN: 9780170371636

For the arm spans:

How: *I will fasten a tape horizontally to the whiteboard with 0 at the left edge. Each student will stand with their back to the board with their arms stretched and the tip of their longest finger on their right hand against the edge. I will measure the distance to the tip of their longest finger on their left hand.*

Managing variation:

- *I will make sure that each person stands with their back to the board, because you get different arm spans if you stand facing the board.*
- *I will make sure that each person stretches their arms horizontally as far as they can in order to ensure consistency.*
- *We will have the same person reading all of the measurements so that the tape measures are read in the same way.*

For each of the following, state what variables you would measure, why, and what units you would use. Then describe how you would measure your variables and how you would manage variation.

1 What is the relationship between people's head circumference and the length of their foot?

Variable 1

What: ______________________________

____

____

Units: ______________

Why: ______________________________

____

How: ______________________________

____

____

____

Managing variation: ______________________________

____

____

____

____

ISBN: 9780170371636

Variable 2

What: ______________________________________

Units: __________________

Why: ______________________________________

How: ______________________________________

Managing variation: ______________________________

2 What is the relationship between the height balls are dropped from and the height of the first bounce?

Variable 1

What: ______________________________________

Units: __________________

Why: ______________________________________

How: ______________________________________

ISBN: 9780170371636

Managing variation: ______________________________

____

____

____

____

Variable 2

What: ______________________________

____

____

Units: ______________

Why: ______________________________

____

How: ______________________________

____

____

____

Managing variation: ______________________________

____

____

____

____

 ISBN: 9780170371636

3 What is the relationship between people's 'before exercise' pulse rates and their 'after exercise' pulse rates?

4 What is the relationship between people's writing speed with their dominant hand and their writing speed with their non-dominant hand?

 ISBN: 9780170371636

Deciding how much data you need and avoiding bias

- You will need to decide **how much data** you will collect in your sample.
- You will need to explain how you tried to **avoid bias**.

A **census** is where we collect data from **every** member of the population. This is usually too expensive, or impossible.

So, when collecting data we nearly always have to take a **sample** from our population. We then use sample data to make **inferences** about the population.

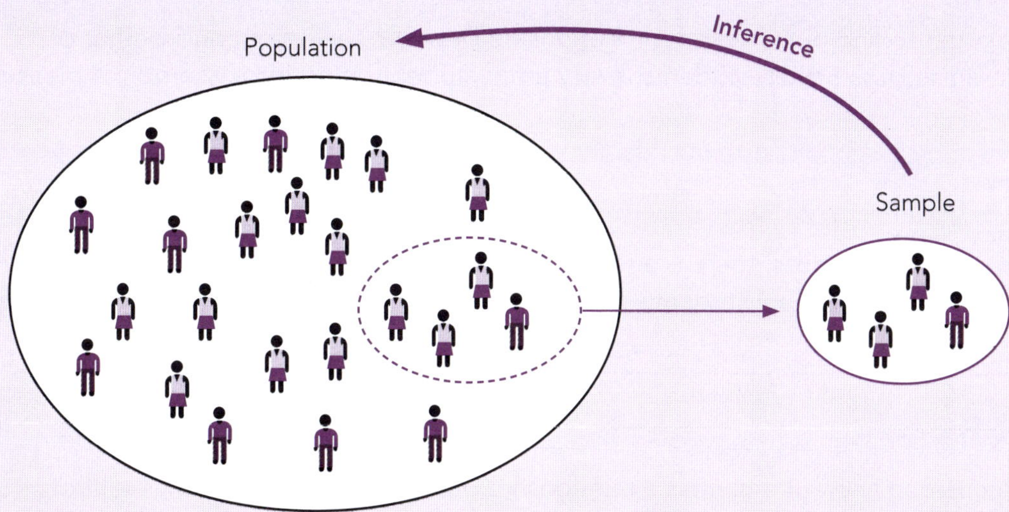

There are two important questions to ask when considering data from a sample:

1 Is the sample big enough?

A sample of 30 is generally considered enough for most purposes. However, the bigger the sample, the more confidence you can have of your findings.

2 Is the data biased?

Bias occurs when some members of the population are more likely than others to be selected for the sample, so the sample does not truly represent the population.

Examples:

- Ringing a radio station, filling in a form in a magazine or going to a website to give feedback: these are called 'self-selected' samples because the subject decides whether they will be sampled. Only those with an interest in the topic of the survey will be in the sample.
- The survey in the mall: this collects data only from those who go to malls and those who have time to answer, and there is an element of self-selection.
- Telephone surveys: are done on land lines, so mostly older people and people who don't often move house are surveyed. These also tend to be self-selected because many people choose not to respond.
- Surveying just your friends.

ISBN: 9780170371636

State whether the following samples might be biased, and why.

1 In order to find out if there is a relationship between the height and weight of Year 10 students, the dean measures the Year 10 girls and Year 10 boys rugby teams.

2 In order to find out if there is a relationship between the height and weight of Paradise High School students, students queuing outside the junior dean's office are measured.

3 In order to find out if there is a relationship between the height and weight of Paradise High School students, every tenth student going into assembly is measured.

4 In order to find out if there is a relationship between the height and weight of Paradise High School students, every tenth student on the school roll is measured.

5 In order to find out if there is a relationship between the height and weight of Paradise High School students, all students with a family name starting with a particular letter of the alphabet are selected.

6 In order to find out if there is a relationship between the height and weight of Paradise High School students, 10 students are selected randomly from each year level.

7 In order to find out if there is a relationship between the height and weight of Paradise High School students, students can go to the school nurse to be weighed and measured.

8 In order to find out if there is a relationship between the time spent reading and English grades of Paradise High School students, students walking past the library at lunchtime are surveyed.

ISBN: 9780170371636

Sampling

In order to avoid bias, there needs to be a system for sampling the population so that **every member of the population has an equal chance of being selected**.

Some systems of sampling

1 **Systematic sampling:** this involves taking every *n*th member of the population.
Examples:
- Selecting every 5th sheep as they pass through a narrow gap.
- Giving every member of the population a number, and then selecting every 10th individual, or those with odd numbers.
- Selecting every 5th car in a car park.
- Selecting every 20th person on the school roll.

2 **Random sampling**
Examples:
- Writing names on equal-sized pieces of paper and drawing them from a hat.
- Writing every number on a ball and using a machine to select the balls, as in Lotto.
- Giving every member of the population a number and using random numbers to decide who will be in the sample.
- Using random numbers to decide who will be sampled from the electoral roll.

How to use random numbers

1 Give all the members of your population a number.
2 Find the 'Random number' button on your calculator.
3 Usually, once in Random mode, if you press the '=' sign you will get a three-digit random number.
4 Decide how big your random number will need to be:
If you have 99 or fewer members, you will need two-digit random numbers.
If you have between 100 and 999 members, you will need three-digit random numbers, etc.
5 Assume there are 60 in your population so you need two-digit random numbers.
This means that you use **only** the first two digits after the decimal point:

Random number	Number selected	Explanation
0.**20**3	20	
0.**58**2	58	
0.**71**1	none	Only 60 in the population, so ignore this random number.
0.0**45**	4	Do not ignore the 0.
0.**00**3	none	Number 0 does not exist.
0.1	10	Zeros at the end are not written, but are understood to be present.
0.**58**9	none	58 has already been selected.

6 Keep going until you have the number required for the sample.

Answer the following.

1 There are 90 students in Year 11. You need to select 30 for your sample.

a Describe two ways in which you could take a systematic sample from these Year 11 students.

i ______________________________

____

ii ______________________________

____

b You have an alphabetical list of the 90 students' names. You numbered these from 1 to 90. The list below shows the first 20 random numbers from your calculator. Complete the table to show which numbers you would select from your list of students, and add any explanations necessary.

Random number	Number selected	Explanation
0.383		
0.974		
0.341		
0.012		
0.688		
0.034		
0.987		
0.493		
0.400		
0.767		
0.522		
0.501		
0.007		
0.83		
0.268		
0.881		
0.322		
0.997		
0.836		
0.08		

ISBN: 9780170371636

2 You want to select a sample of 20 of your school's 86 staff.

a Describe two ways in which you could take a systematic sample from these 86 staff.

i ____________________

____

ii ____________________

____

b You have an alphabetical list of the 86 staff names. You numbered these from 1 to 86. The list below shows the first 20 random numbers from your calculator. Complete the table to show which numbers you would select from your list of staff, and add any explanations necessary.

Random number	Number selected	Explanation
0.868		
0.661		
0.468		
0.783		
0.479		
0.987		
0.269		
0.866		
0.091		
0.879		
0.188		
0.044		
0.263		
0.25		
0.017		
0.464		
0.009		
0.332		
0.7		
0.08		

ISBN: 9780170371636

Describing and comparing features

1 Direction of the relationship

Positive relationship:

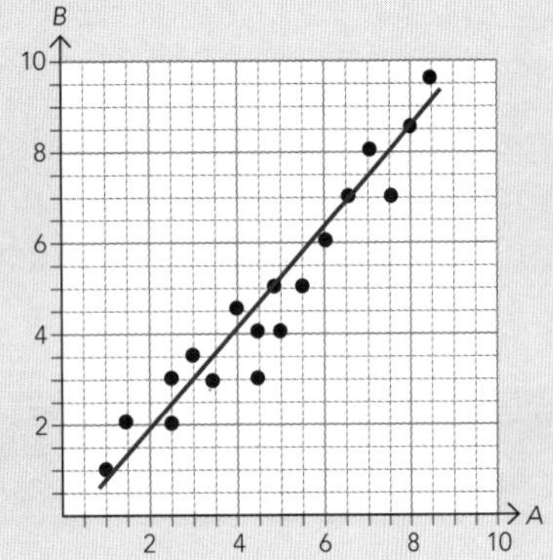

As A increases, B **increases**.

Negative relationship:

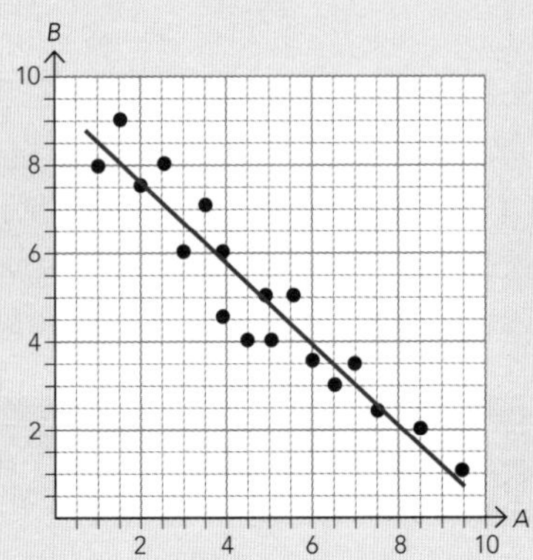

As A increases, B **decreases**.

Example 1: The graph shows the ages (to the nearest year) of men and women in 26 couples.

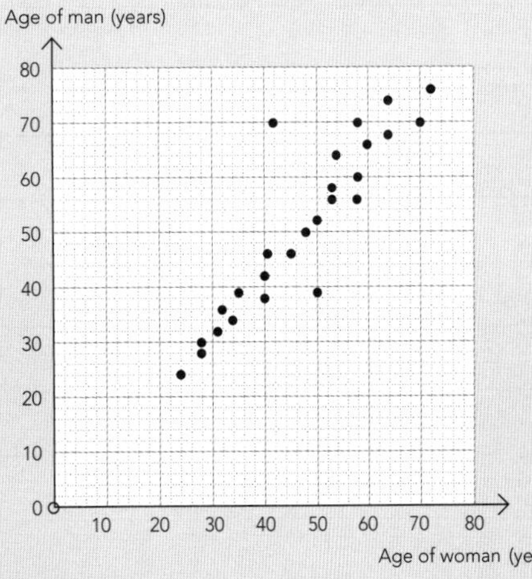

I notice that there is a positive relationship between **the age of the woman** and the **age of the man** in a couple.

This means that as the **age of the woman** increases, the **age of the man** also tends to increase.

Example 2: The graph shows the relationship between age in years of children and the time taken to complete a puzzle (min).

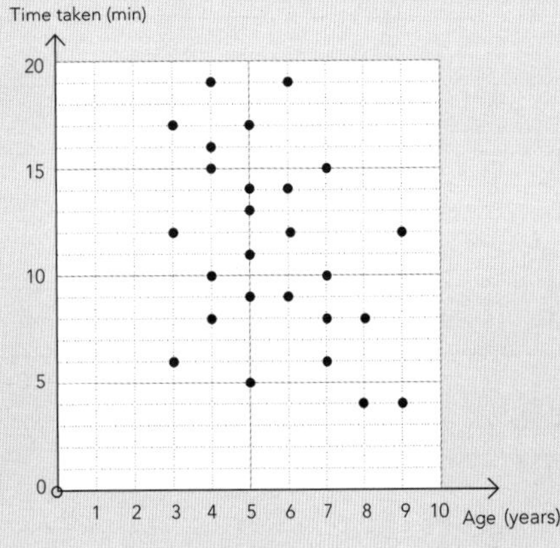

I notice that there is a negative relationship between the **age of the child** and the **time taken to complete the puzzle**.

This means that as the **age of the child** increases, the **time taken to complete the puzzle** tends to decrease.

ISBN: 9780170371636

Write statements about the direction of the relationships below:

1 The graph shows the relationship between head circumference and best long-jump distance for some Year 11 students.

Distance jumped (cm)

Head circumference (cm)

I notice that there is a ____________________ relationship between ____________________ ____________________ and ____________________ ____________________.

This means that as ____________________ increases, ____________________ tends to increase/decrease.

2 The graph shows the relationship between age of cars in years and price of a range of second-hand cars.

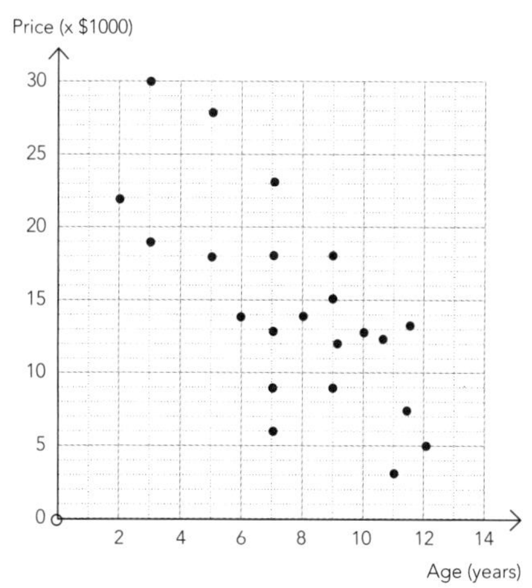

I notice that there is a ____________________ relationship between ____________________ ____________________ and ____________________ ____________________.

This means that as ____________________ increases, ____________________ tends to increase/decrease.

3 The graph shows the relationship between price and the number of each model of car sold in a year.

Number

Price (x $1000)

____

____

____

____

____

____

____

____

ISBN: 9780170371636

4 The graph shows the relationship between the price of a new car and its weight (kg).

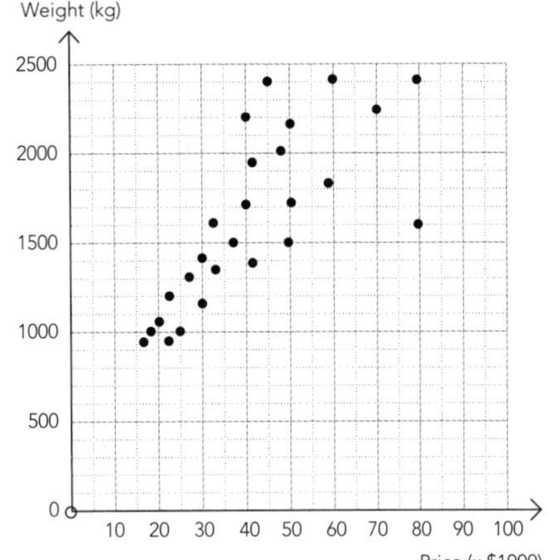

5 The graph shows the relationship between students' height (cm) and their accuracy in a basketball goal-shooting test (%).

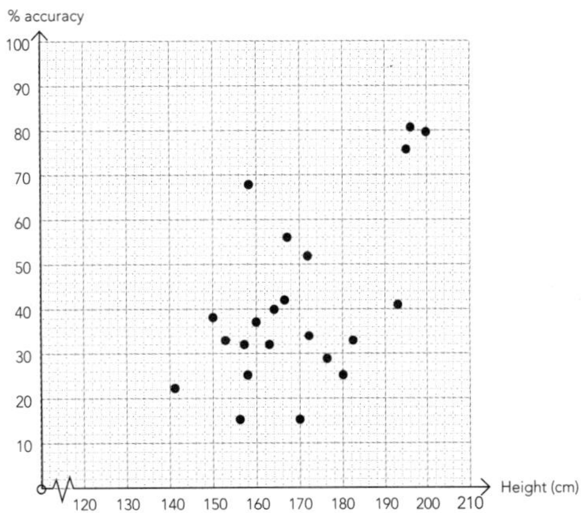

6 The graph shows the relationship between age in years and the time taken to complete a puzzle (min).

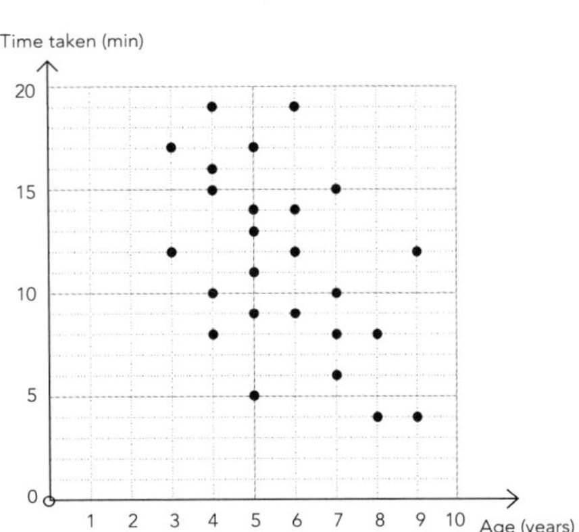

ISBN: 9780170371636

7 The graph shows the relationship between average income per person and average life expectancy for a range of countries.

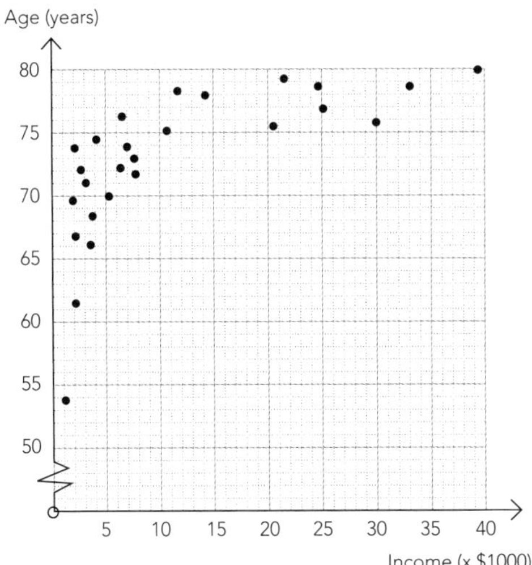

8 The graph shows the relationship between the number of hours per week spent playing video games and the number of credits earned at Level 1 by a group of students.

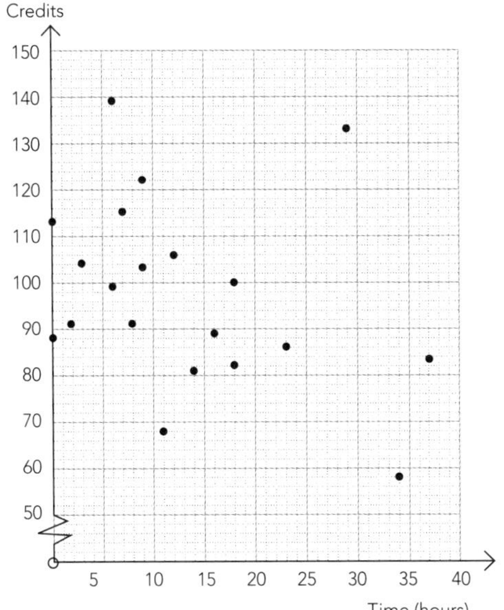

2 Linear and non-linear relationships

- Usually the line of best fit is a **straight** line: the relationship is **linear**.
  Example: The relationship between students' arm span (cm) and height (cm).

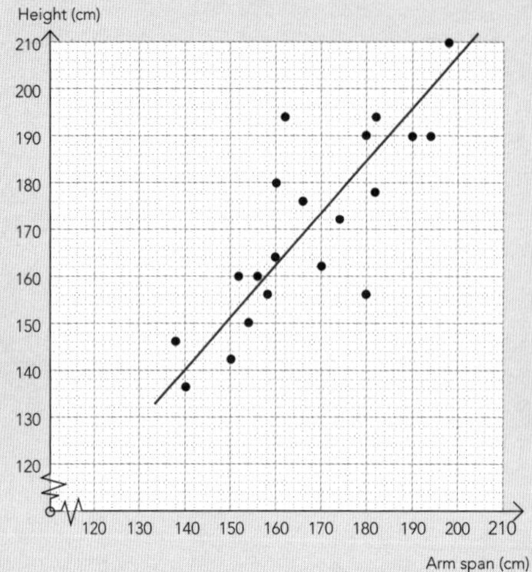

- Sometimes the line of best fit is a **curve**: the relationship is **non-linear**.
  Example: The graph shows the relationship between the number of each model of car sold in a year and its price.

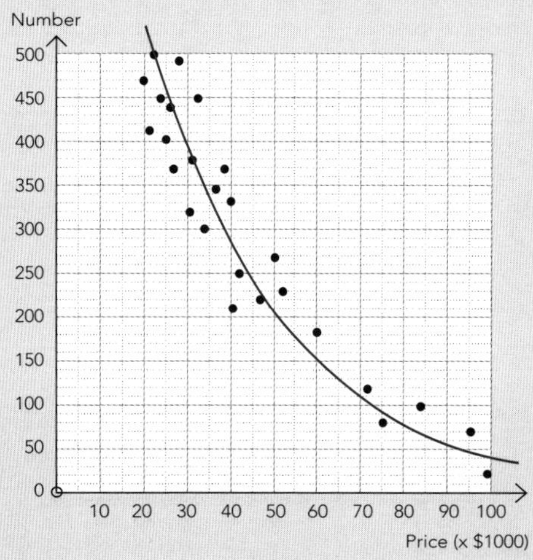

Decide whether each of the following is linear or non-linear. Circle the correct answer.

1 The graph shows the relationship between the price of a new car and its weight (kg).

Linear **Non-linear**

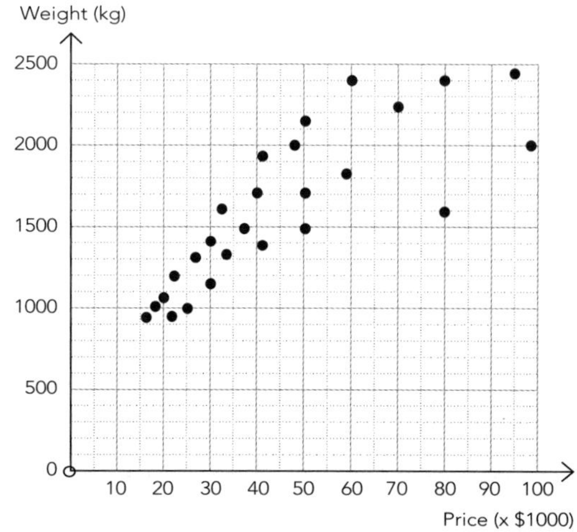

ISBN: 9780170371636

2 The graph shows the relationship between head circumference and best long-jump distance for some Year 11 students.

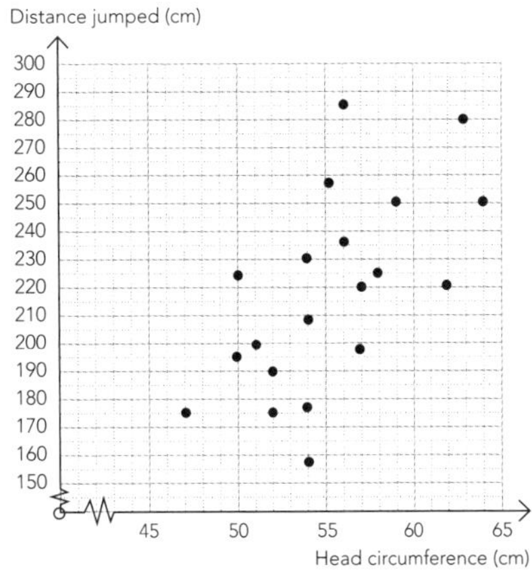

Linear **Non-linear**

3 The graph shows the relationship between age in years and the time taken to complete a puzzle (min).

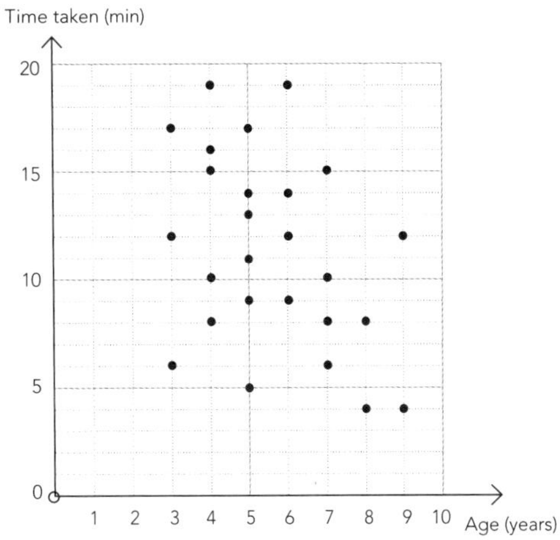

Linear **Non-linear**

4 The graph shows the relationship between price and the number of each model of car sold in a year.

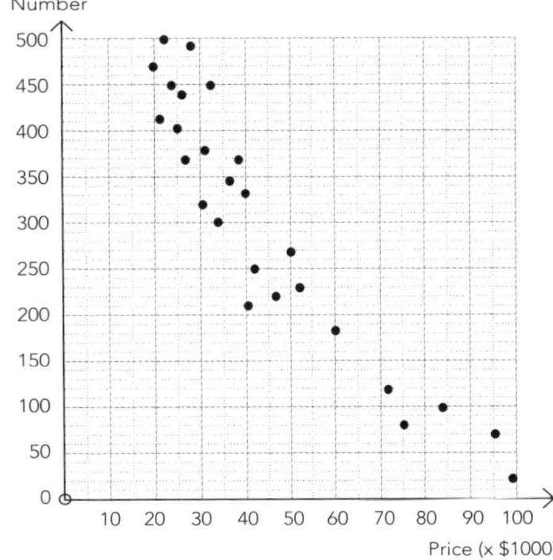

Linear **Non-linear**

5 The graph shows the relationship between age in years and price of a range of second-hand cars.

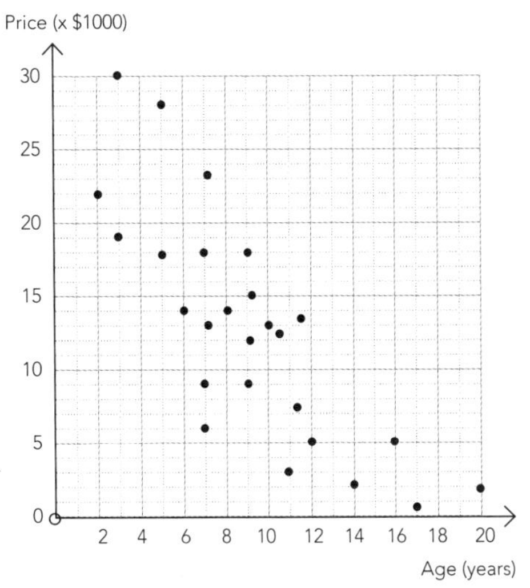

Linear **Non-linear**

ISBN: 9780170371636

3 Strength of the relationship

- You need to describe how close the points are to the line of best fit.
- You need to state how strong the relationship is.
- You need to say what effect that has on your confidence in your conclusions.

Strong:

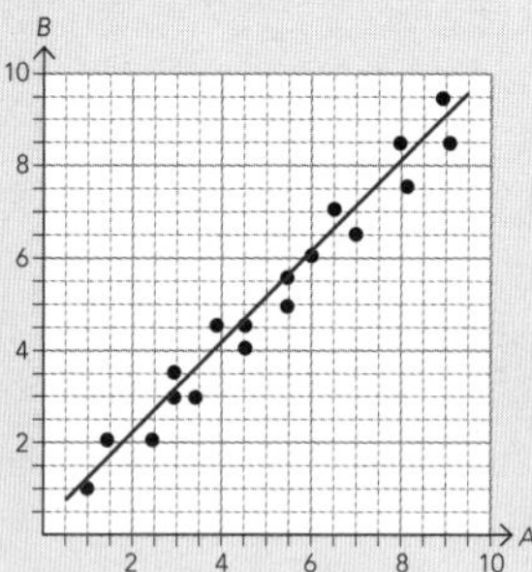

I notice that:	most points are close to the line of best fit.
This means that:	the relationship between A and B is **strong**.

Moderately strong:

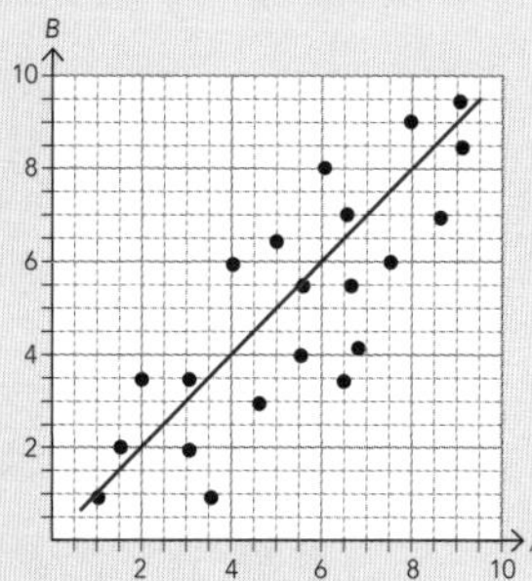

I notice that:	most points are fairly close to the line of best fit.
This means that:	the relationship between A and B is **moderately strong**.

Moderately weak:

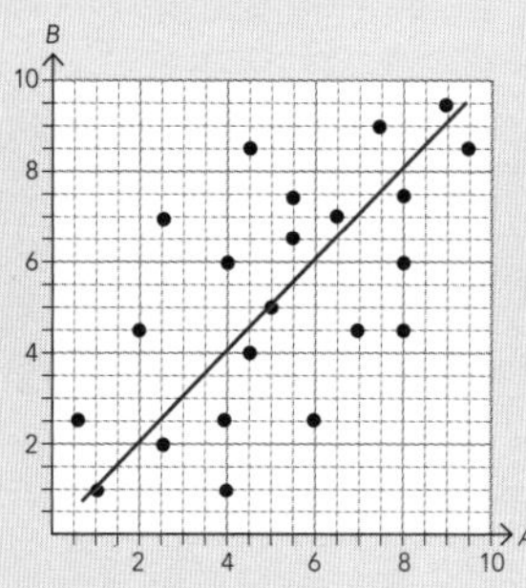

I notice that:	some points are quite a long way from the line of best fit.
This means that:	the relationship between A and B is **moderately weak**.

Weak:

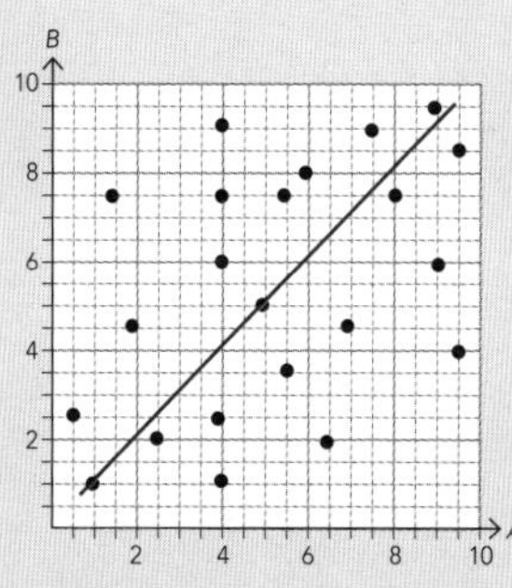

I notice that:	lots of points are quite a long way from the line of best fit.
This means that:	the relationship between A and B is **weak**.

ISBN: 9780170371636

No relationship:

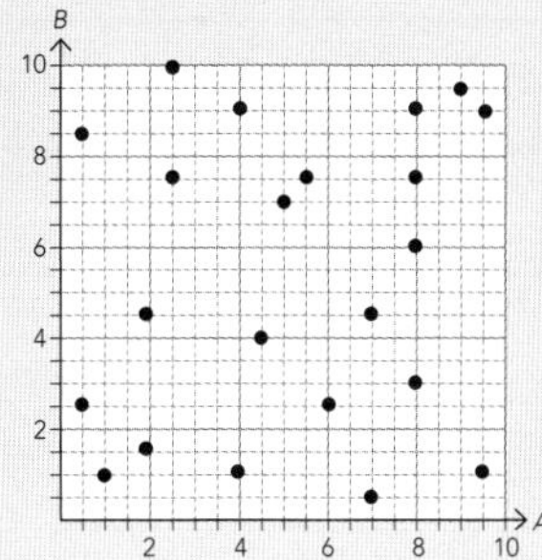

I notice that: the points are so scattered that I cannot draw a line of best fit.

This means that: there is **no** relationship between A and B.

Describe how close the points are to the line of best fit, then state how strong the relationship is and how this affects your confidence in your conclusion.

1 The graph shows the relationship between head circumference and best long-jump distance for some Year 11 students.

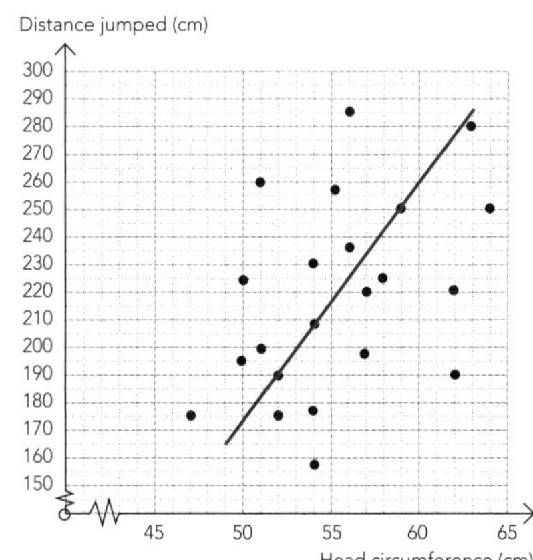

I notice that: ______________________________

This means that: ______________________________

2 The graph shows the relationship between the number of hours per week spent playing video games and the number of credits earned at Level 1 by a group of students.

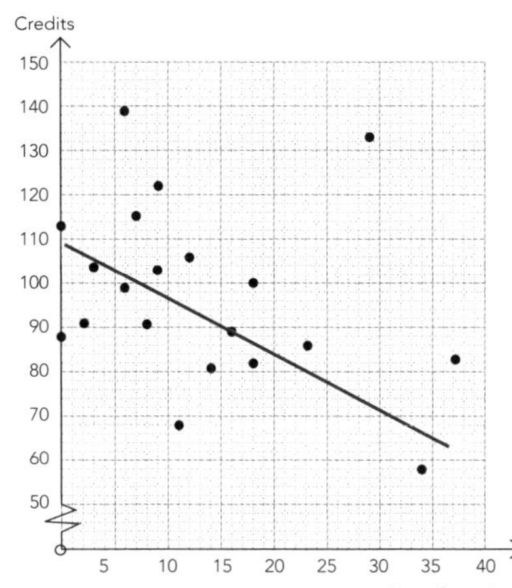

I notice that: ______________________________

This means that: ______________________________

ISBN: 9780170371636

3 The graph shows the relationship between car weight (kg) and fuel consumption (L/100 km).

Fuel consumption (L/100 km)

30
25
20
15
10
5
0
5 10 15 20 25
Weight (x 100kg)

I notice that: ______________________

This means that: ______________________

4 The graph shows the relationship between price and the number of each model of car sold in a year.

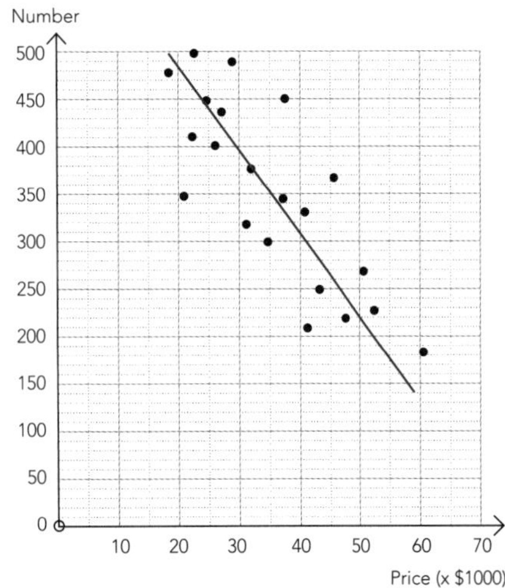

I notice that: ______________________

This means that: ______________________

5 The graph shows the relationship between the weight of a new car and its price.

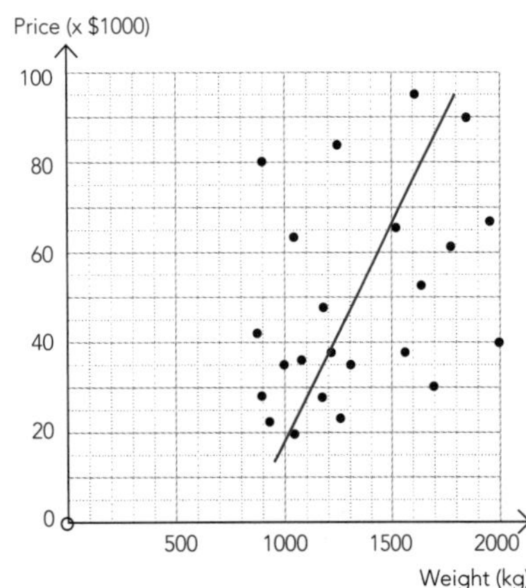

I notice that: ______________________

This means that: ______________________

 ISBN: 9780170371636

4 Uneven scatter

- Sometimes the degree of scatter changes across the graph.
- This means that the relationship is stronger in some parts than others.

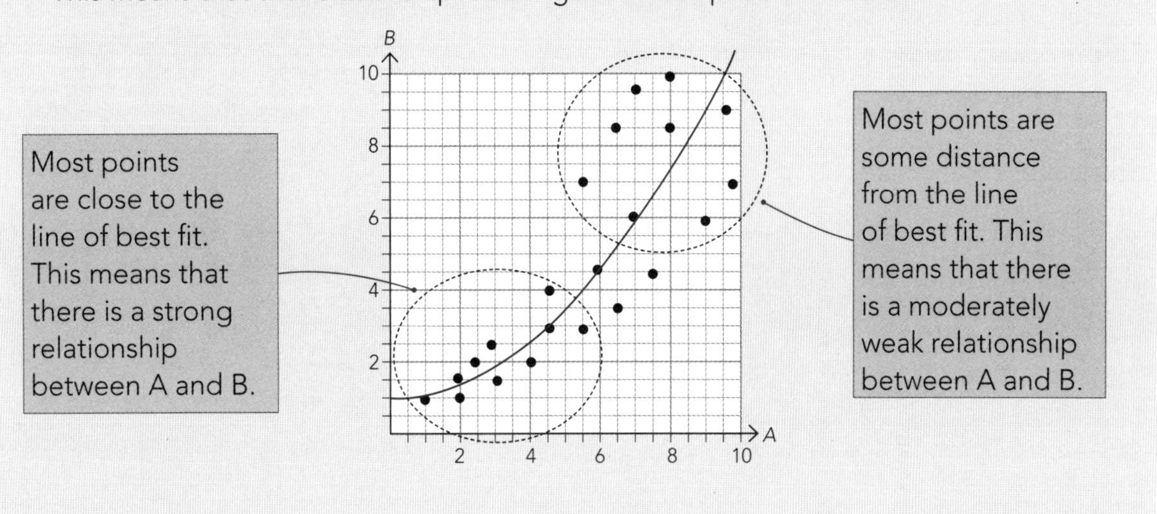

Describe how close the points are to the line of best fit, then state how strong the relationship is and how this affects your confidence in your conclusion.

1 The graph shows the relationship between the price of a new car and its weight (kg).

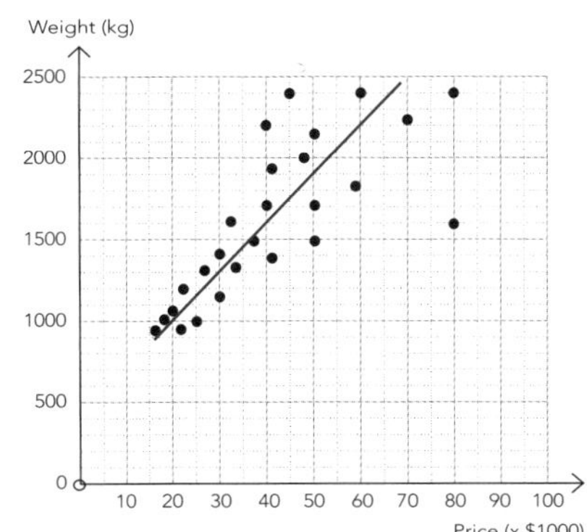

I notice that: ______________________

____

____

This means that: ______________________

____

____

2 The graph shows the relationship between car weight (kg) and fuel consumption (L/100 km).

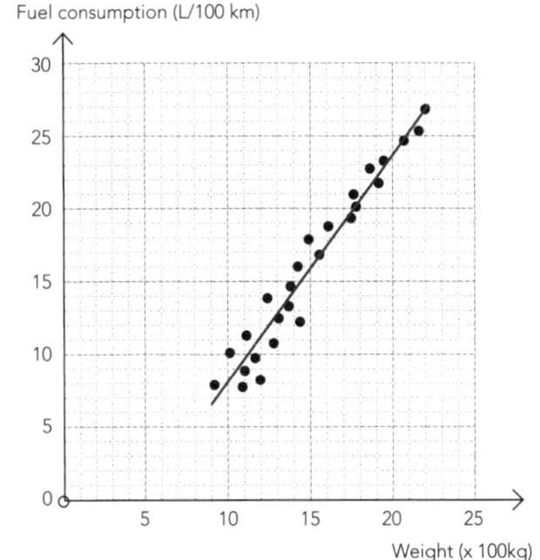

I notice that: ______________________

____

____

This means that: ______________________

____

____

5 Unusual features

Unusual features may be unusual points or clusters of points.

- These are points that lie some distance from most of the other points.
- Sometimes these are valid points.
- Otherwise they may be a result of measurement error, recording error, or reversed co-ordinates.

Examples:

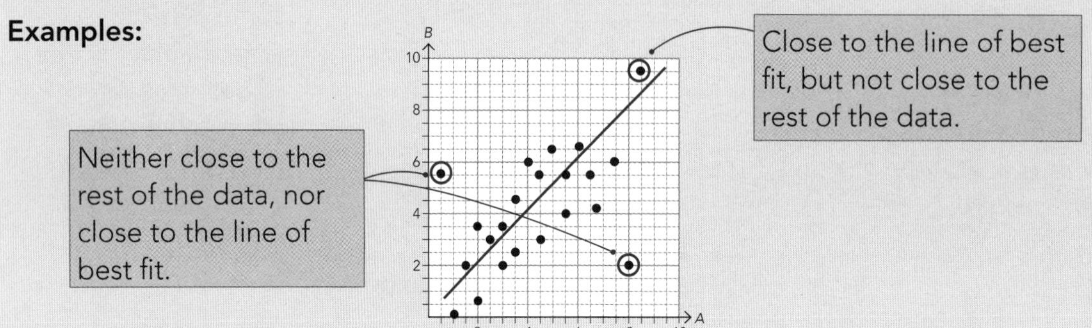

Unusual clusters of points

These are small groups of data that lie some distance from other points.
You should always discuss these, and try to find an explanation.

These may be:

1 In line with the line of best fit.

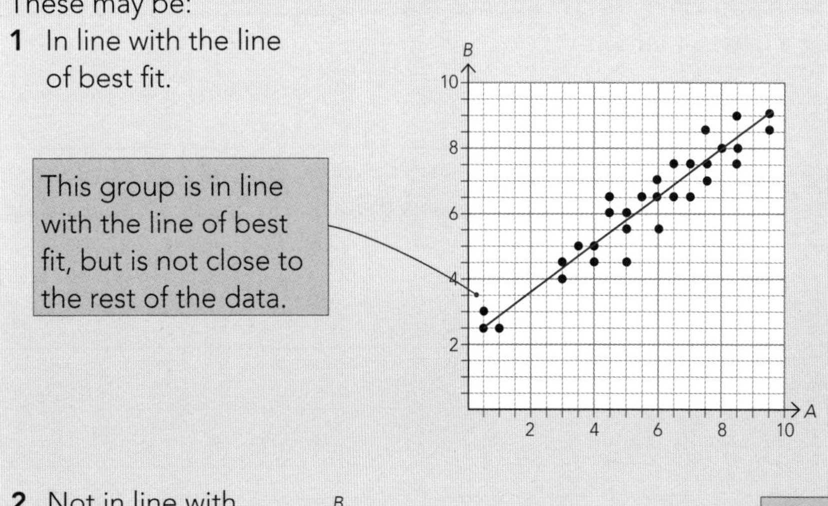

2 Not in line with the line of best fit.

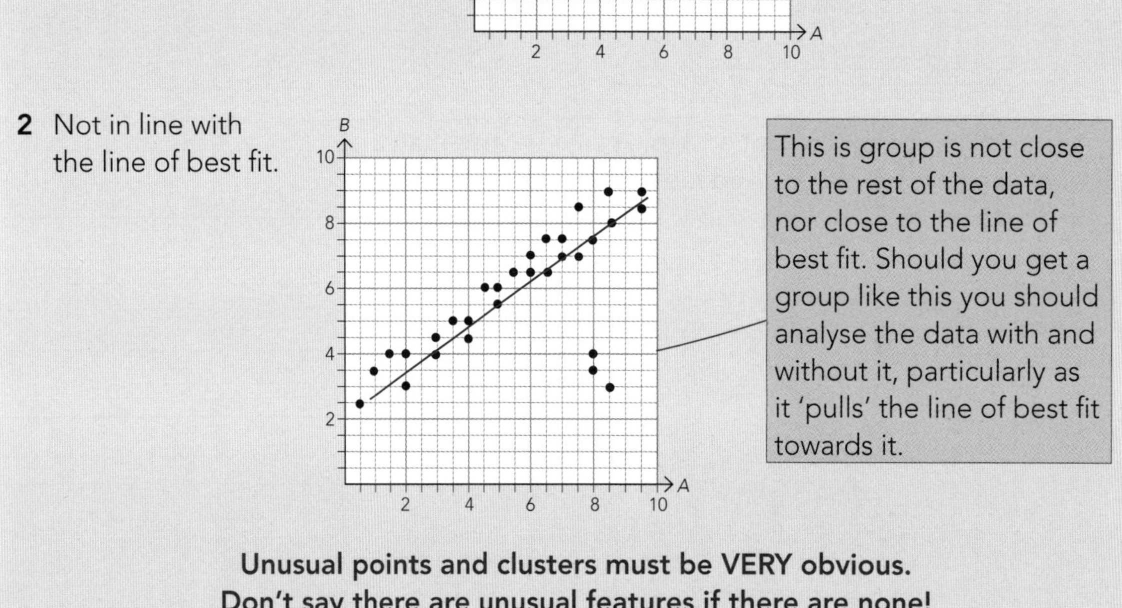

Unusual points and clusters must be VERY obvious.
Don't say there are unusual features if there are none!

ISBN: 9780170371636

Example: Consider the arm spans and heights of a group of Year 11 students.

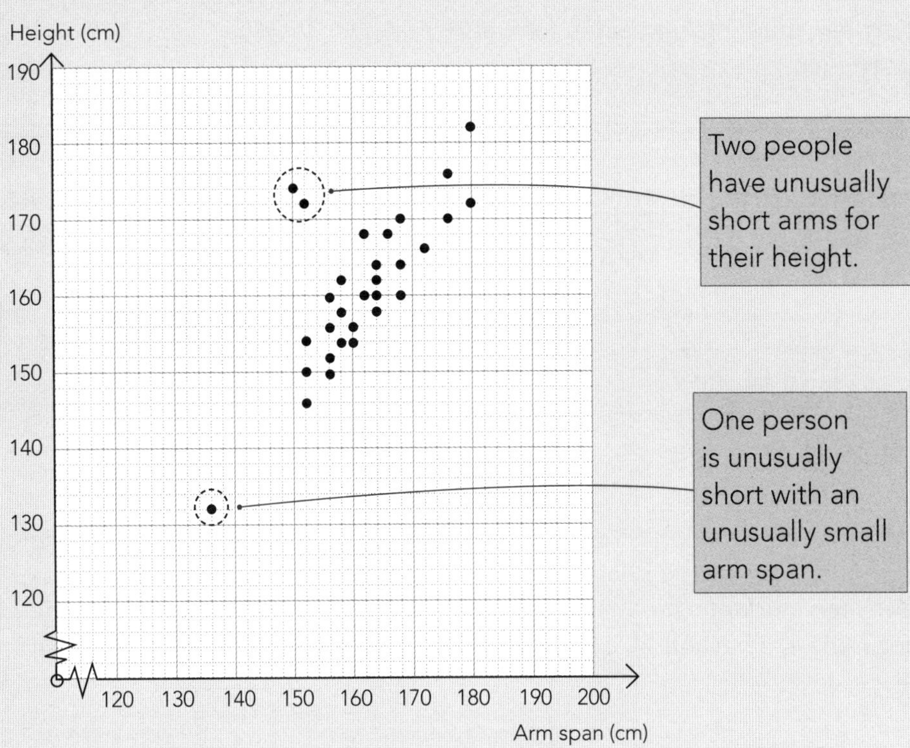

What do you see?

I notice that there is a cluster of smaller values and one larger value that are significant distances from most of the data.

Why might we have these clusters or unusual features? Link to the context if possible.

What does this mean for the sample?

This means that there are two students who, given their height (172 cm and 174 cm), have unexpectedly small arm spans (152 cm and 150 cm) compared with the majority of the sampled students. Their arm spans and heights are not in the same proportion as the rest of the students because their points are a long way from the line of best fit. They are both unusual so perhaps they are twins.

There is also one student who is much shorter than most (132 cm), and with a smaller arm span (136 cm). Although this student was much smaller than the others, their arm span is in proportion to their height because their point would be close to the line of best fit.

This person could be much younger than the other students, or perhaps he or she was a dwarf.

ISBN: 9780170371636

Describe any unusual features in the following (not all graphs have them).

1 The graph shows the relationship between students' arm span (cm) and height (cm).

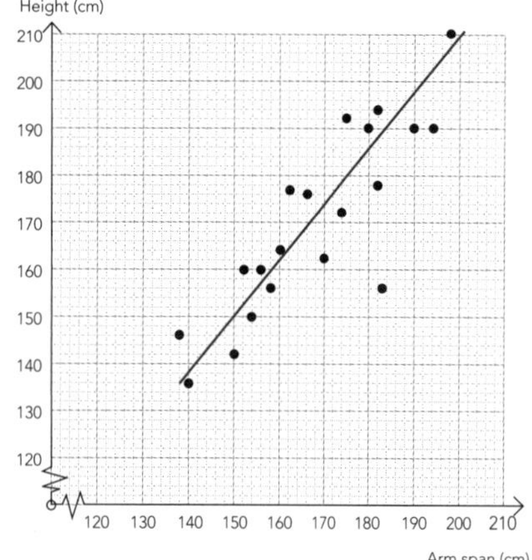

2 The graph shows the relationship between students' height (cm) and their accuracy in a basketball goal-shooting test (%).

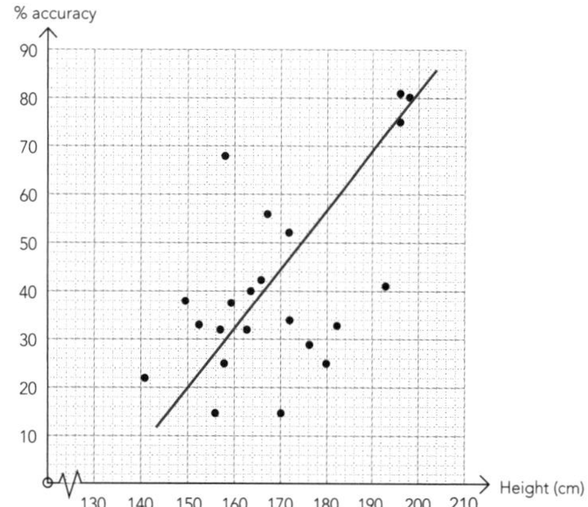

3 The graph shows the relationship between the number of hours per week spent playing video games and the number of credits earned at Level 1 by a group of students.

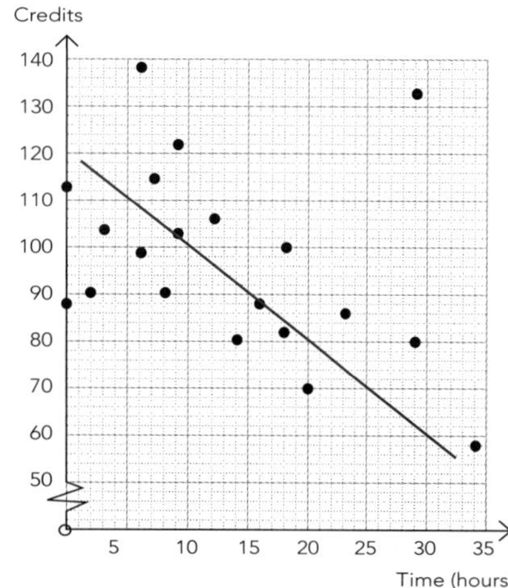

 ISBN: 9780170371636

4 The graph shows the relationship between the price of a new car and its weight (kg).

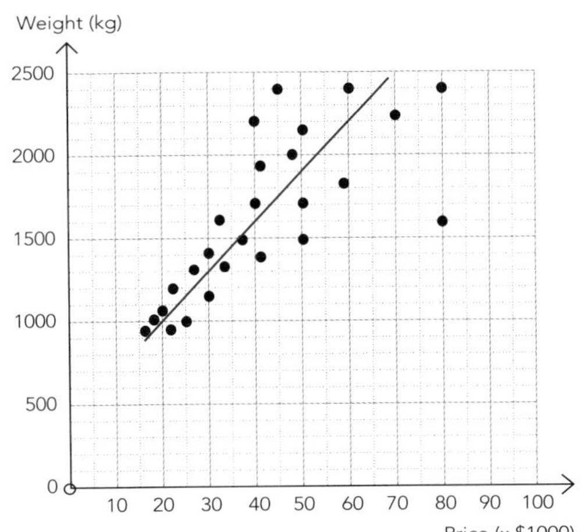

5 The graph shows the relationship between age in years, and price of a range of second-hand cars.

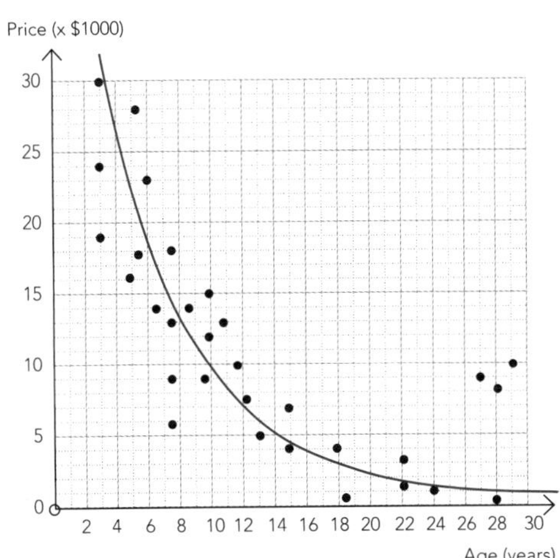

6 The graph shows the relationship between the number of each model sold in a year and its price.

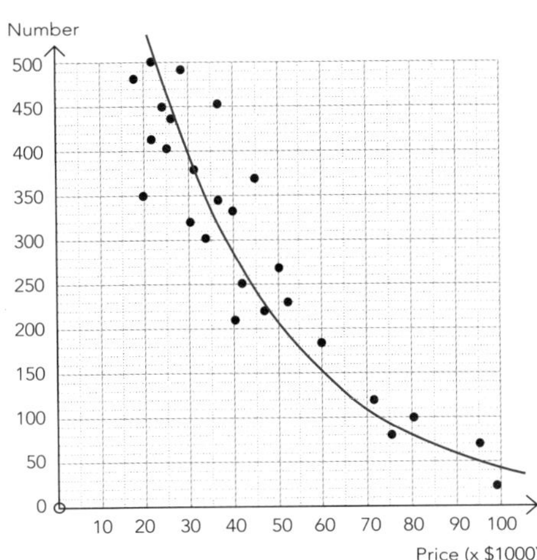

6 Cleaning the data

- This means removing pieces of data that are **obvious** mistakes.
- If possible, check the piece of data, re-measure the subject, check the units, etc.

Examples:

1 Heights of Year 11 students in cm:
156, 149, 181, 194, **276**, 187, 143, 173.

Remove this piece: not a feasible height.

2 Marks out of 20:
18, 14, 9, 11, 16, 13, **27**, 20, 16, 15.

Remove this piece: more than 20.

3 Prices of cars:
$3400, **45**, $10,999, $7499, $13,300, $7699, $5499, $8700.

Remove this piece: an unlikely price for a car and no $ sign.

- **Always** state in your report whether or not you have removed pieces of data, and justify your decisions.

Do not remove unusual points unless you have a good reason for doing so.

State which points you would remove (if any) from the following data sets, and why.

1 Heights of Year 11 students (cm):

149, 184, 159, 173, 165, 1.65, 142, 179, 151, 151, 166, 161, 171, 148, 159, 150

---

2 Heights of Year 11 students (cm):

169, 171, 189, 201, 169, 176, 152, 188, 171, 169, 184, 149, 159, 178, 157, 173

---

3 Heights of Year 11 students (cm):

159, 162, 159, 146, 177, 170, 182, 277, 141, 178, 150, 144, 175, 167, 150, 163

ISBN: 9780170371636

4 Heights of Year 11 students (cm):

169, 151.5, 193, 201, 169, 178, 160, 180, 167, 166, 182, 158, 149, 167, 157, 170

5 The following shows some of the data collected during an investigation into use of cellphones by Year 11 students at Paradise High School.

Gender	Number of texts sent	Number of texts received	Number of calls made	Number of calls received
M	15	17	5	2
F	225	192	7	13
M	43	5000	0	1
M	-21	28	3	6
M	67	61	1	11
F	92	99	2.3	9
F	144	137	17 min	22 min

State which pieces of data you would remove, and why.

ISBN: 9780170371636

Putting it together

You will need to do the following.

- Write down the **question** you have been given to investigate.
- Discuss the variables:
  - State **what** variables you will use.
  - State what **units** you will use.
  - Explain **why** you have chosen each variable.

  Describe in detail **how** you will measure your chosen variables.
- Discuss how you will manage **variation**.
- Discuss your **sampling**:
  - **How much data** did you collect in your sample and **why**?
  - How did you try to **avoid bias**?
- Describe **how** the data was recorded:

  Examples: I wrote the information in a table on a piece of paper.
I entered the information onto an Excel spreadsheet.
  - Give details of the **column headings** in your table or spreadsheet.
  - **Name** the people who performed each role.
- **Display** the data:
  - Draw a **scatter** plot.
  - Draw a **line** of best fit.
  - Explain **why** a scatter plot was the best way of displaying the data.
- **Describe** the relationship in terms of the context and whether it is:
  - positive/negative, including a **relationship statement**: 'As variable *A* increases, variable *B* increases/decreases.'
  - strong/weak
  - evenly scattered or not
  - linear/non-linear, etc
  - other observations.
- Describe any **unusual features** and whether you have **cleaned** the data.
- Write a conclusion, which **MUST** refer back to your question.
- Discuss how you could **improve** the investigation if you did it again.

ISBN: 9780170371636

Annotated example

Question

I am interested in how long a person can stand on one foot with their eyes closed. I would like to know if there is a relationship between how long a Year 11 student at Paradise High School can stand on their dominant foot compared with their non-dominant foot.

Variables

My variable will be the length of time that each person can stand on each foot with their eyes shut. We will measure the time in seconds. ✓ what ✓ units

We will tell each subject to shut their eyes and lift one foot from the floor and keep it off the floor for as long as possible. We will use a stopwatch and record how long it is before the second foot touches the floor, and whether they were standing on their right or left foot. After they have balanced on each foot, we will ask each person whether their right or left foot is dominant (normally the first foot used when a person starts walking). ✓ how

Variation

In order to avoid variation we will have the same people giving the instructions, using the stopwatch, and recording the data throughout the data collection. ✓

We will make everybody do this in bare feet, because the type of shoe worn could affect how long people can balance. ✓

We will also take people to a quiet room because having classmates talking and calling out would probably affect how long people can balance. ✓

We will not ask about which foot is dominant until after we have done the timing, so people don't know why we are doing this. If they did know they might favour one foot or the other. ✓

must have at least one

Sampling

We will record the balance times for the 30 students in our Maths class. A sample of 30 is generally considered enough to give a clear result. We have not sampled randomly from Paradise High School's Year 11, which would have been ideal and ensured there was no bias. However, we could see no reason why using one Maths class would give biased results for a variable like balance.

✓ how much data ✓ why ✓ avoiding bias

How the data was recorded

Heidi gave the instructions to each person, Pania did the timing, and I wrote down the results in a table with the following headings: ✓ roles identified

✓ column headings

	Time (s)		
Name	**Right foot**	**Left foot**	**Dominant foot**
Joe Blogs	145	118	R

Data display

A scatter graph was the best way to display this data because it let me show both variables at the same time. It also allows a line of best fit to be drawn, which is helpful in discussing the data.

✓ why

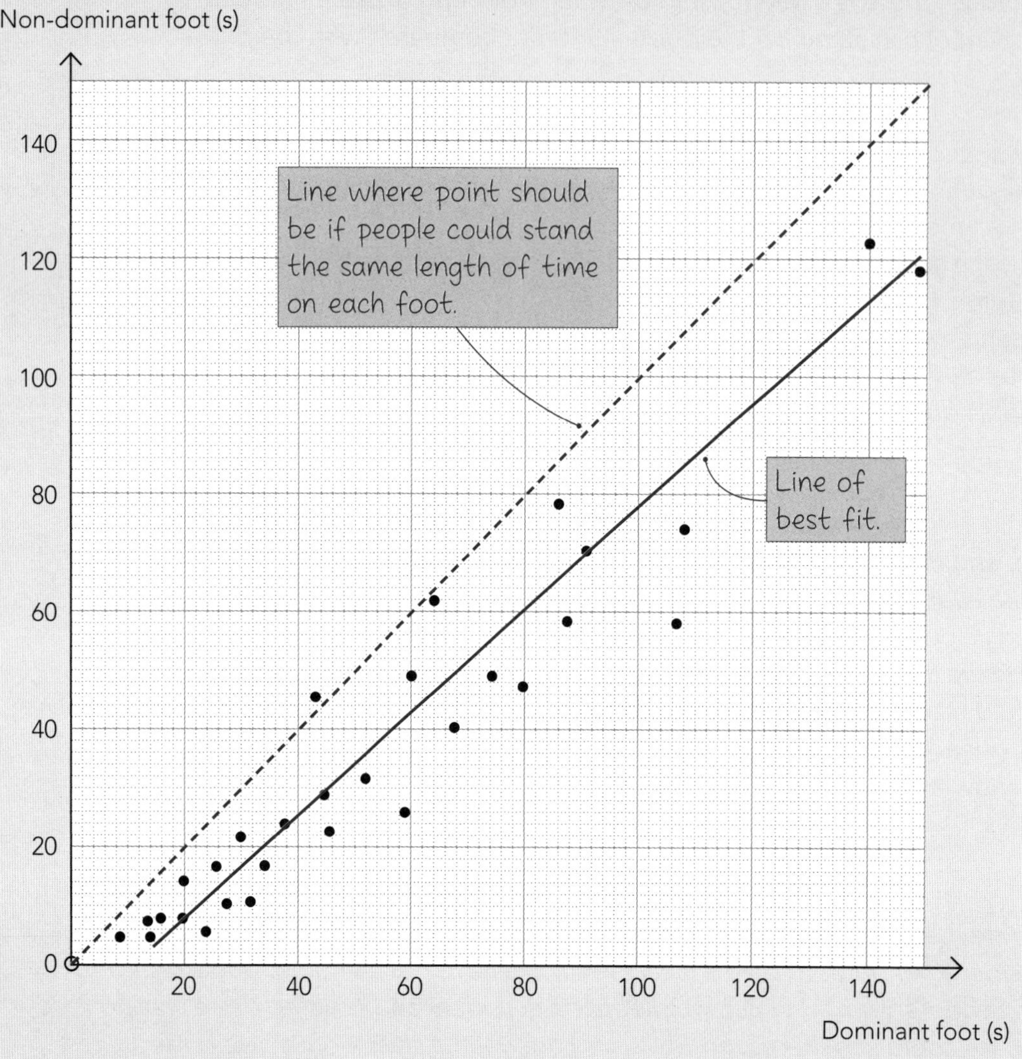

✓ scatter plot

✓ line of best fit

Description of the relationship

There is a positive relationship between the length of time a Year 11 student can stand on their dominant foot when they have their eyes closed, and how long they can stand on their non-dominant foot. This means that as the time a person can stand on their dominant foot with their eyes closed increases, the time they can stand on their non-dominant foot tends to increase also. This does not surprise me, because how long people can stand on one foot will mostly depend on how good their balance is.

✓ relationship statement

The relationship is quite strong because most points are fairly close to the line of best fit. However, I notice that the points are closer to the line of best fit for people who could stand on one foot for less than 40 seconds. Above 40 seconds, the data is more scattered. This means that the relationship is stronger for people who could stand on one foot for less than 40 seconds.

✓ strength

ISBN: 9780170371636

The relationship is linear, because I could draw a straight line of best fit through the data, and the points were fairly evenly scattered on both sides.

I notice that people can usually stand for longer on their dominant foot than on their non-dominant foot. I have drawn a dotted line on the graph, which is where the points should be if people could stand for the same length of time on each foot.

Only the person who stood for 43 seconds on their dominant foot and 45 seconds on their non-dominant foot stood for longer on their non-dominant foot. Everybody else was able to balance for longer on their dominant foot.

✓ linear

✓ other observations

Unusual features

There are two points that are unusual: one person stood for 140 seconds on their dominant foot and 123 seconds on the other foot; the second stood for 148 seconds on their dominant foot and 118 seconds on the other. While their points were close to the line of best fit, they both had much better balance than anybody else. The nearest value was 80 seconds. Possibly these students are ballerinas, where balancing on one foot would be very important.

I did not remove any pieces of data.

Conclusion

I have found that there is quite a strong relationship between how long a Year 11 student at Paradise High School can stand on their dominant foot compared with their non-dominant foot. Students who can stand for a long time on one foot tend to be able to stand for a long time on the other foot. I have also found that students tend to be able to stand for longer on their dominant foot than their non-dominant foot.

✓ refers to question

Improvements

I could have more confidence in my results if I took a bigger sample. I would also have more confidence in my results if I sampled from the whole Year 11, and not just my maths class. It would also be interesting to know if this relationship is the same for all ages of students and adults. I would also like to know if the results are similar when subjects have their eyes open.

ISBN: 9780170371636

Reflections

These are some reflective questions that you could consider answering in your report.

Did my sample fairly represent the population in my question?

Was my sample big enough?

Did I try to prevent variation in my measurements?

Did I make any mistakes in my investigation?

Are all my statements true?

Do my conclusions seem reasonable?

Did I answer my question?

If I did this again, how could I improve my investigation?

ISBN: 9780170371636

Practice tasks

Practice task one

Question

I would like to know if there is a relationship between how far a Year 11 student from Paradise High School can hop using their dominant foot compared with their non-dominant foot. This data is from a sample of Year 11 students from Paradise High School.

Data

	Distance (cm)		
Name	**Right foot**	**Left foot**	**Dominant foot**
Karl	68	59	R
Annie	79	75	R
Jess	24	51	R
Tane	87	85	R
Heong	90	75	R
Mick	102	100	R
Stacey	41	35	R
Rosie	69	63	R
Adam	77	81	L
Aroha	56	55	R
Chloe	49	50	R
Simone	35	30	R
Zoe	41	38	R
Liam	62	57	R
Daniel	92	96	L
Ian	56	53	R
Kim	72	78	L
Yi Ming	34	32	R
Zack	67	64	R
Henry	93	85	R
Pete	84	71	R
Lavini	72	67	R
Sam	66	62	R
George	70	57	R
Meg	59	55	R
Joe	53	50	R
Selini	82	81	R
Will	65	51	R
Nick	49	48	R
Fred	13	18	L

ISBN: 9780170371636

Variables

Variation

Sampling

  ISBN: 9780170371636

How the data was recorded

Data display

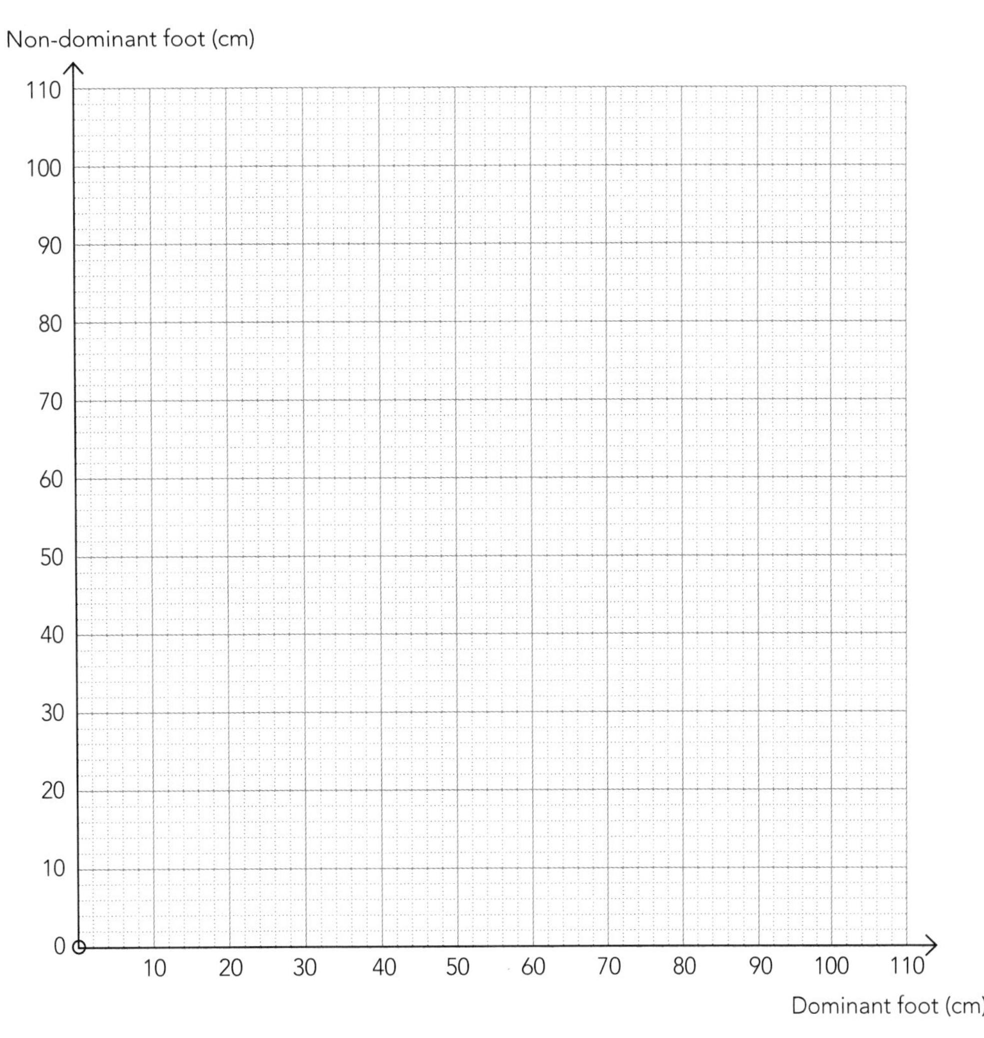

Description of the relationship

Unusual features

Conclusion

Improvements

ISBN: 9780170371636

Practice task two

Question

I would like to know if there is a relationship between height and how long it takes a Year 11 boy at Paradise High School to run 100 m. The data below is from a sample of Year 11 boys at Paradise High School.

Data

Name	Height (cm)	Time to run 100 m (s)
Karl	160	19.2
Joe	168	14.5
Jason	159	18.1
Tane	153	18.2
Heong	151	20.9
Mick	160	14.0
Stephen	157	20.1
Rob	156	14.9
Adam	163	15.2
Arana	162	14.6
Chris	168	13.1
Simon	166	14.8
Zane	168	14.0
Liam	165	16.1
Daniel	155	23.8
Ian	169	13.5
Kevin	171	13.0
Whetu	169	14.3
Zack	165	13.8
Henry	177	23.4
Pete	162	14.2
Liam	164	15.0
Sam	163	14.2
George	176	11.2
Mike	162	16.7
John	149	19.8
Selini	165	15.1
Will	174	14.3
Nick	159	21.2
Fred	166	15.9

ISBN: 9780170371636

Variables

Variation

Sampling

 ISBN: 9780170371636

How the data was recorded

Data display

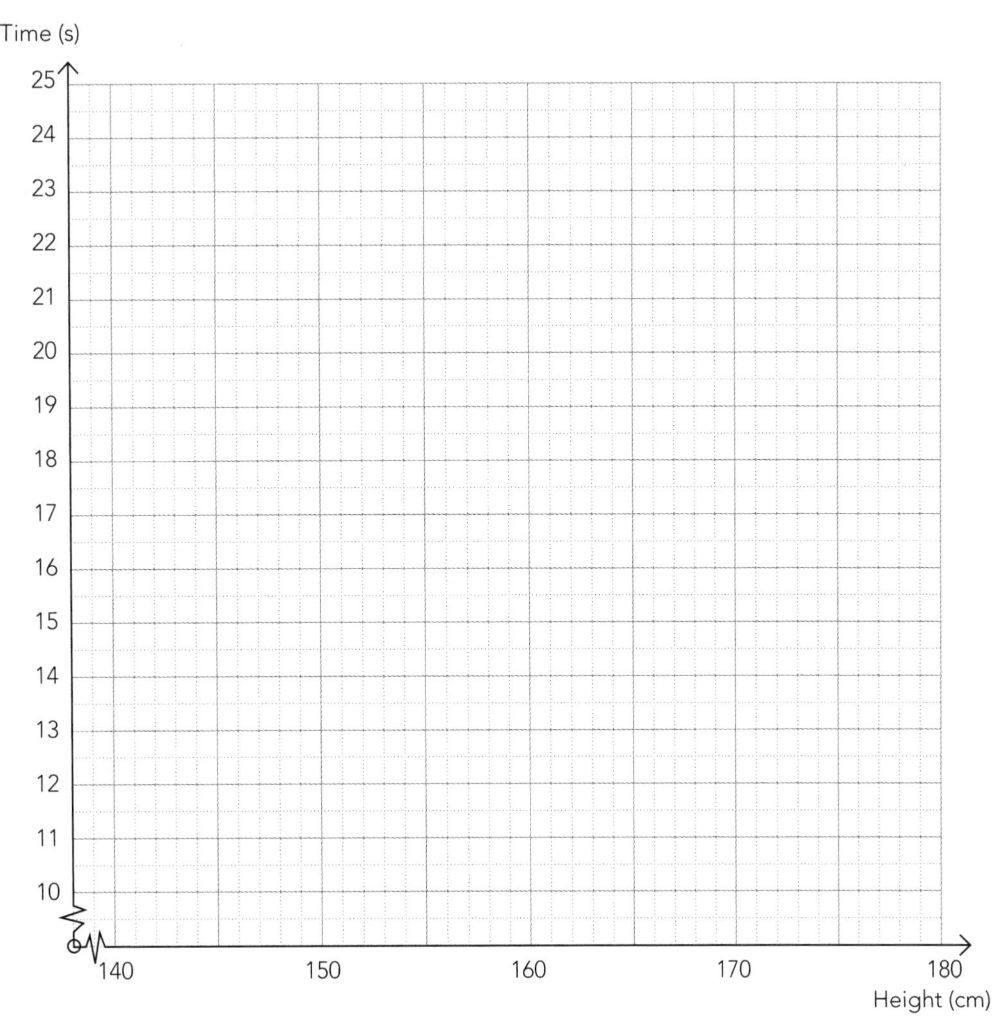

ISBN: 9780170371636

Description of the relationship

Unusual features

Conclusion

Improvements

 ISBN: 9780170371636

Practice task three

Question

I would like to know whether there is a relationship between mouth size (height of the gap between front teeth when the mouth is wide open) and how long it takes for a Year 11 student at Paradise High School to eat a dry Weet-Bix and then drink a glass of milk. This data is from a sample of Year 11 students from Paradise High School.

Data

Name	Mouth size (mm)	Time taken (s)
Karl	42	137
Annie	53	237
Jess	50	151
Tane	40	279
Heong	37	244
Mick	49	240
Stacey	43	262
Rosie	51	221
Adam	55	172
Aroha	42	128
Chloe	37	335
Simone	39	219
Zoe	41	668
Liam	46	167
Daniel	47	169
Ian	33	412
Kim	57	225
Yi Ming	45	215
Zack	41	241
Heny	41	355
Pete	48	440
Lavini	44	321
Sam	45	375
George	40	215
Meg	42	255
Joe	41	124
Selini	43	108
Will	54	293
Nick	39	372
Fred	47	383

ISBN: 9780170371636

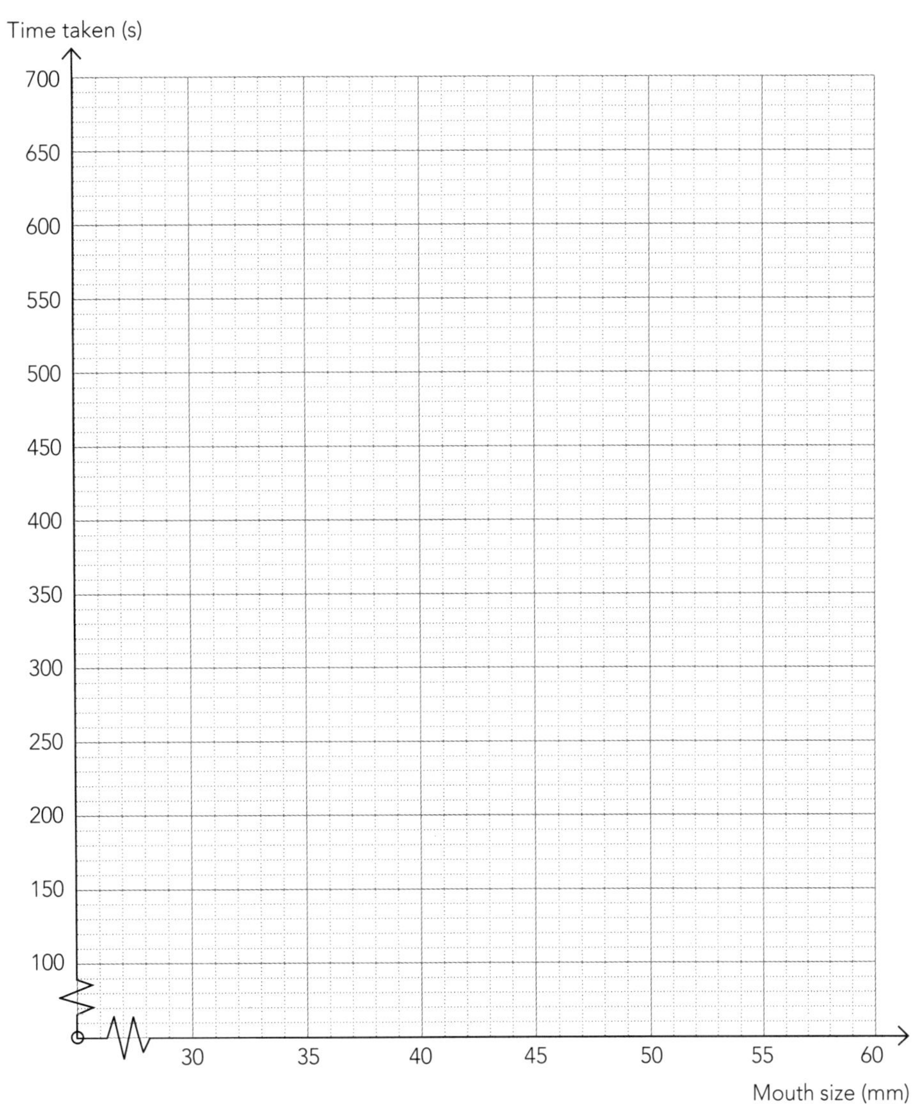

ISBN: 9780170371636

ISBN: 9780170371636

Answers

Probabilities are rounded to a maximum of 4 dp. Professional judgement should apply.

Variables (pp. 6–7)

Variable	Type of variable	Suitable for a bivariate study?
Favourite colour	Descriptive	✗
Foot length	Continuous	✓
Number of pens in pencil case	Discrete	✗
Head circumference	Continuous	✓
Shoe size	Discrete	✗
Number of pets	Discrete	✗
Number of text messages on phone	Discrete	Possibly, depending on range
Favourite ice-cream flavour	Descriptive	✗
Height	Continuous	✓
Area of your bedroom	Continuous	✓
Distance travelled to school	Continuous	✓
Eye colour	Descriptive	✗
Arm span	Continuous	✓
Make of calculator	Descriptive	✗
Age	Continuous	✓
Time taken to eat a dry Weet-Bix	Continuous	✓
Amount spent in canteen last week	Discrete	✗
Weight of school bag	Continuous	✓

Any combinations of the continuous variables except, possibly, the distance travelled to school because it is unlikely that there will be a relationship between distance travelled to school and the other continuous variables. For example:

Height	and	Foot length
Height	and	Head circumference
Age	and	Foot length
Arm span	and	Height
Age	and	Head circumference

Display of data (pp. 8–18)

Creating scatter plots (pp. 9–11)

1

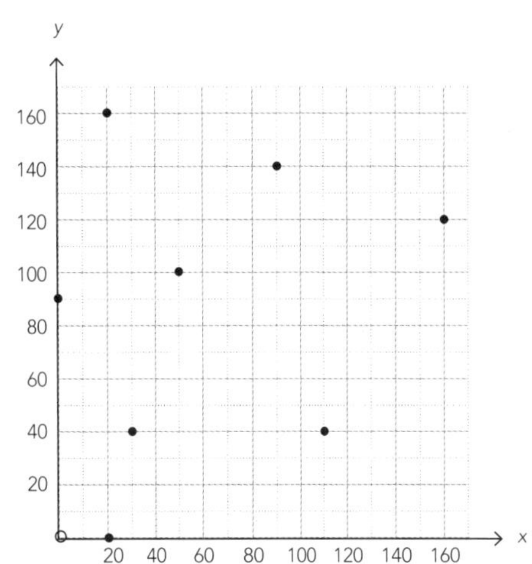

2

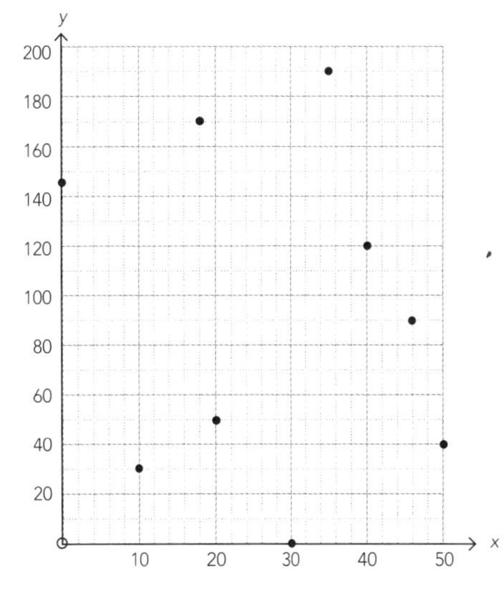

3

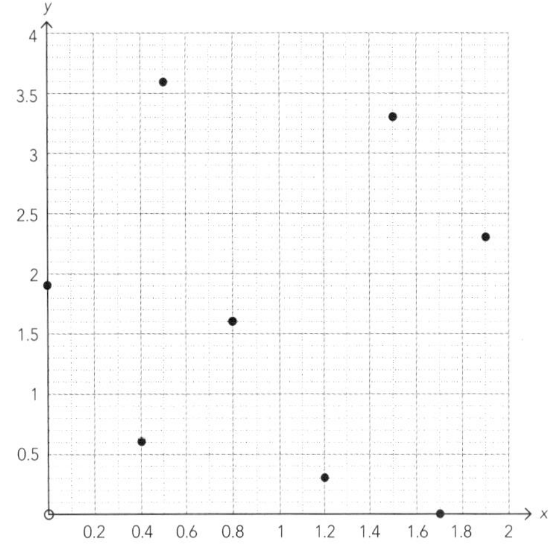

ISBN: 9780170371636

4

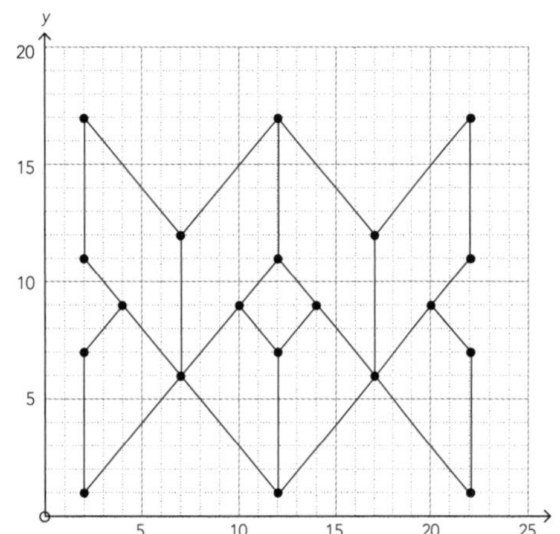

Reading scatter plots (pp. 12–13)

1 a (158, 156)
 b (198, 210)
 c (140, 136)
 d (190, 190)
 e (180, 156)

2 a (157, 32)
 b (156, 15)
 c (193, 41)
 d 197 cm
 e 33%

3 a (12, 106)
 b (29, 133)
 c (11, 68)
 d 81 credits
 e 9 hours

Drawing a line of best fit (pp. 14–15)

1

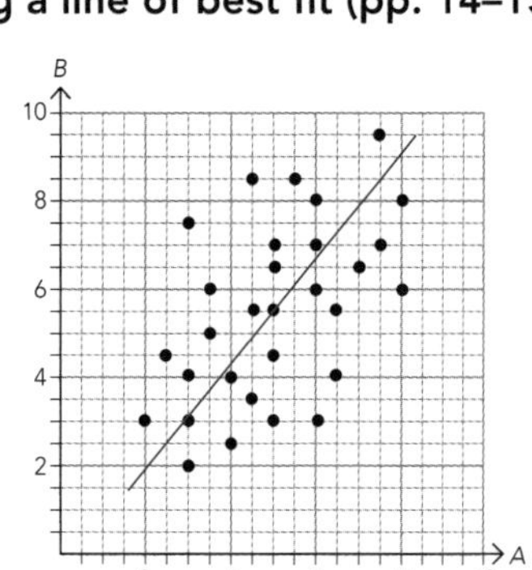

2

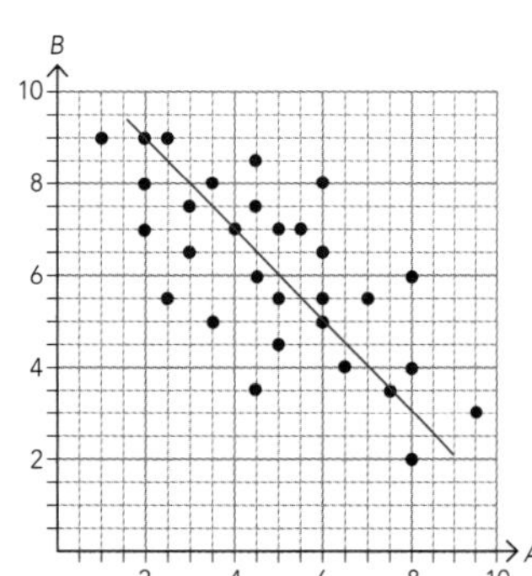

3

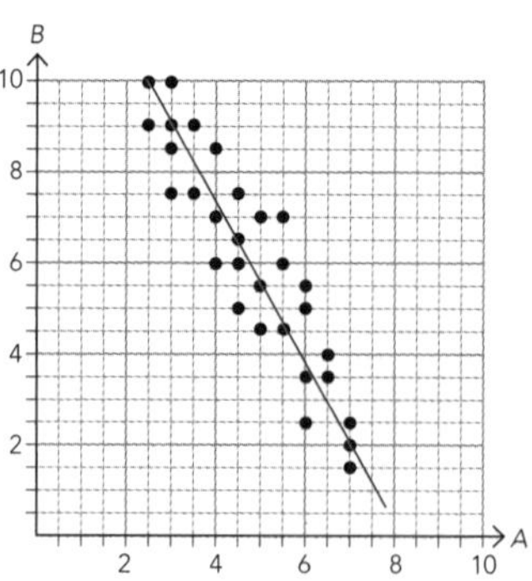

4

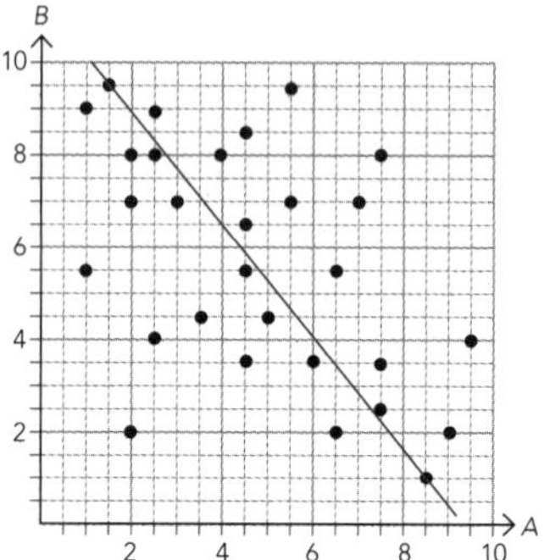

5

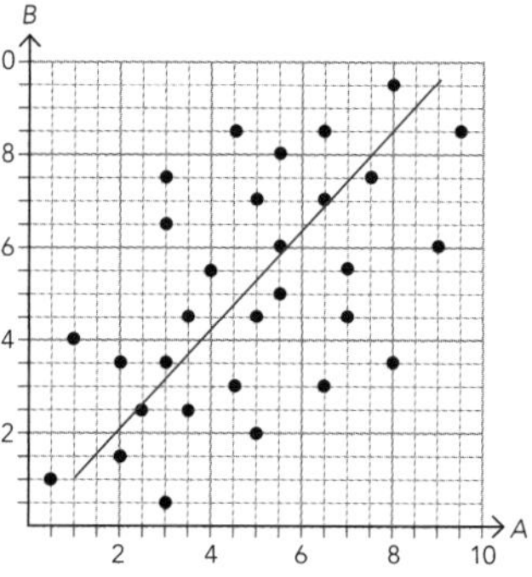

6

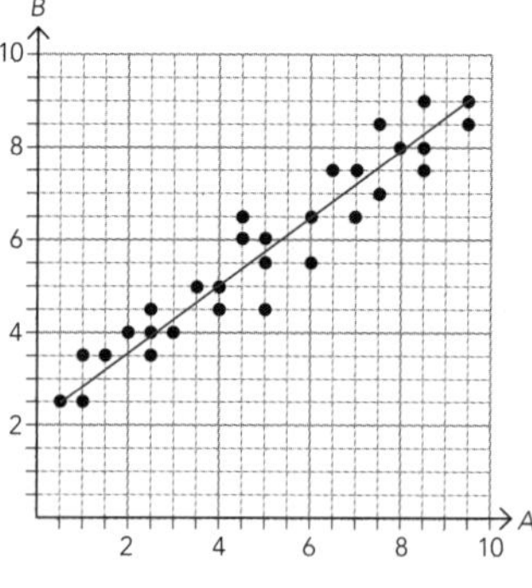

Estimating values from a line of best fit (pp. 16–18)

1 a 180 cm b 60 cm
 c 285 cm

2 a 46% b 192 cm
 c 146 cm

3 a 13.5 years b 10 kg
 c 9.5 years

4 a 26 days
 b 900 mm
 c 26 days because the points are much closer to the line for birds under 600 mm.

 ISBN: 9780170371636

Planning an investigation (pp. 19–29)

1 **Variable 1**

What: The distance around each person's head.
Units: cm
Why: This will give me a measure of their head circumference.
How: I will use a tape measure and wrap it horizontally around the biggest part of each person's head. I will read the length on the tape.
Managing variation: For each person, I will make sure that I record the biggest part of their head. This may involve taking several measurements at different positions and selecting the biggest, because different people have different head shapes.
I will make sure that each person is sitting upright and looking straight ahead, and that I measure horizontally. Otherwise the measurements would not be consistent.
We will get the same person to do all the measuring.
We will make sure that the tape measure doesn't overlap.

Variable 2

What: The maximum length of each person's foot, from the tip of their longest toe to back edge of their heel.
Units: cm
Why: This will give us a measure of their overall foot length.
How: We will stick a tape measure to the floor with 0 at the wall and the tape measure at right angles to the wall. We will ask each person to put their heel against the wall, and we will record where the end of their longest toe is.
Managing variation: We will ask everybody to do this with a bare foot, so we get just their foot length without variable-sized shoes or socks.
We will measure everybody's right foot.
By asking them to put their heel hard against the wall, we should get reasonably accurate foot lengths.
We will have the same person doing all the measuring.

2 **Variable 1**

What: The height above the floor of the top of the ball before it is dropped.
Units: cm
Why: I will use the height above the floor of the top of the ball because I want to measure to the top of the ball during the first bounce, and it is important to measure to the same part of the ball for both variables.
How: We will stick a measuring tape to the wall with 0 at the floor. We will hold the ball a few mm out from the tape measure and record the height of the top of the ball before we drop it.
Managing variation: We will use the same ball throughout, because different balls, even if they are all tennis balls, would bounce differently. We will have the same person drop the ball each time so that it is dropped in the same way. Some people might bounce rather than drop the ball. We will make sure we measure the height of the ball by reading the tape horizontally in order to get an accurate height, and not from above or below.

Variable 2

What: The maximum height of the top of the ball during its first bounce.
Units: cm
Why: I have chosen to measure the maximum height of the top of the ball because I think that will be easier to judge.
How: We will estimate the maximum height of the top of the ball during its first bounce by watching the ball carefully, and read off the tape measure stuck to the wall.
Managing variation: The ball will be moving fast so this will not be easy. In order to be as accurate as possible, we will take three measurements at each height, and take the average of all three.

3 My first variable will be the number of heartbeats in one minute, taken before exercise starts. The units are beats per minute. This will give a measure of pulse rate. The subject will sit in a chair. One person will hold their wrist and count the number of beats. A second person will use a stopwatch and tell the other person when to start counting, and to stop after exactly one minute.

In order to reduce variation, we will make sure that the same person does all the counting, and the same person does all the timing. We will also make sure that all subjects rested for at least 20 minutes before they have their resting pulse measured. Otherwise their first pulse might not be a

ISBN: 9780170371636

truly resting pulse. They will all be sitting in a chair when their pulse is measured, because whether a person is sitting, standing or lying down might make a difference to their pulse rate.

My second variable will be the number of heartbeats in one minute, taken immediately after exercise stops.

In order to reduce variation we will give all subjects the same amount of exercise to do, and will make sure that we take their second pulse rate as soon as they stop exercising. Otherwise their pulse might drop before we got the second measurement done. Otherwise it will be measured in the same way that we measured the first pulse rate.

4 **Variable 1**
What: The length of time it takes to write 'The quick brown fox jumps over the lazy dog' with their dominant hand.
Units: s (seconds)
Why: The length of time taken to write a particular amount gives a measure of speed, because everyone is writing the same words.
How: I will give each person a piece of paper, a pen and a card with 'The quick brown fox jumps over the lazy dog' typed on it. I will use a stopwatch to measure how long they take to copy the sentence.
Managing variation: I will make sure everybody has the same sort of pen and paper, because some pens and pencils are harder to write with, so writing is slower. I will make sure the same person does all the timing, because some people might be slower in pressing the stopwatch at the start and finish. I will make sure that everybody is sitting at a desk and in similar conditions.

Variable 2
This will be exactly the same as Variable 1, and will be measured in the same way, except that each person will be using their non-dominant hand.

Deciding how much data you need and avoiding bias (pp. 25–26)

1 This would definitely give biased results because rugby players probably tend to be taller and heavier than non rugby-playing students.

2 This would definitely give biased results because students queuing outside the junior dean's office are likely to be juniors and therefore smaller than the average student at Paradise High School.

3 This would probably give unbiased results, but students who bunk assembly would not be measured. If the bunkers tended to be fatter, thinner, taller, heavier, etc. then the sample could be biased.

4 This should give an unbiased sample.

5 This might give an unbiased sample, but there is the possibility of selecting more or fewer of one nationality, e.g. if they chose L, C or W, more Chinese students might be chosen. Also, families of students might be chosen.

6 This should give an unbiased sample, as long as the numbers at each level are similar. If, for instance, there were fewer students in Year 13 than at other levels, Year 13 students would be over-represented in the sample.

7 This would give a biased sample. The students who go to the school nurse may not be typical of the average student.

8 This would definitely give a biased sample, because students who are near the library are likely to be better readers than the average student.

Sampling (pp. 27–29)

1 **a** i Get an alphabetical list of the 90 students, and select every third person.

ii Take every third Year 11 student as they file out of assembly.

b

Random number	Number selected	Explanation
0.**38**3	38	
0.**97**4	none	Only 90 in the population, so ignore this random number.
0.**34**1	34	
0.**01**2	1	Do not ignore the 0.
0.**68**8	68	
0.**03**4	3	Do not ignore the 0.
0.**98**7	none	Only 90 in the population, so ignore this random number.

 ISBN: 9780170371636

Random number	Number selected	Explanation
0.493	49	
0.400	40	
0.767	76	
0.522	52	
0.501	50	
0.007	none	Number 0 does not exist.
0.83	83	
0.268	26	
0.881	88	
0.322	32	
0.997	none	Only 90 in the population, so ignore this random number.
0.836	none	83 is already selected.
0.08	8	

2 a i Get an alphabetical list of all the staff, and select every fourth until you have the required 20.

ii Take every fourth person in the staff photograph.

b

Random number	Number selected	Explanation
0.868	86	
0.661	66	
0.468	46	
0.783	78	
0.479	47	
0.987	none	Only 86 in the population, so ignore this random number.
0.269	26	
0.866	none	86 is already selected.
0.091	9	Do not ignore the 0.
0.879	none	Only 86 in the population, so ignore this random number.
0.188	18	

Random number	Number selected	Explanation
0.044	4	Do not ignore the 0.
0.263	none	26 is already selected.
0.25	25	
0.017	1	Do not ignore the 0.
0.464	none	46 is already selected.
0.009	none	Number 0 does not exist.
0.332	33	
0.7	70	Zeros understood to be after the 7.
0.08	8	Do not ignore the 0.

Describing and comparing features (pp. 30–45)

1 Direction of the relationship (pp. 30–33)

1 I notice that there is a positive relationship between head circumference and maximum long-jump distance.
This means that as head circumference increases, maximum long jump distance tends to increase.

2 I notice that there is a negative relationship between age and price of second-hand cars.
This means that as age increases, the price of second-hand cars tends to decrease.

3 I notice that there is a negative relationship between the price of a new car and the number sold.
This means that as the price of a new car increases, the number sold tends to decrease.

4 I notice that there is a positive relationship between the price of a new car and its weight.
This means that as the price of a new car increases, its weight tends to increase.

5 There is a positive relationship between a student's height and their accuracy in a basketball shooting test.
This means that as height increases, their accuracy in a basketball shooting test tends to increase.

6 There is a negative relationship between age and time taken to complete a puzzle. This means that as age increases, the time taken to complete a puzzle tends to decrease.

7 There is a <u>positive</u> relationship between the average income per person and the average life expectancy of a country.
This means that as the <u>average income per person</u> increases, the <u>average life expectancy in a country</u> also tends to increase.

8 There is a <u>negative</u> relationship between the number of hours spent playing video games each week and the number of credits earned by students at Level 1.
This means that as the <u>number of hours spent playing video games each week</u> increases, <u>the number of credits earned by students at Level 1</u> tends to decrease.

2 Linear and non-linear relationships (pp. 34–35)

1 Non-linear
2 Linear
3 Linear
4 Non-linear
5 Non-linear

3 Strength of the relationship (pp. 36–38)

1 I notice that some points are quite a long way from the line of best fit. This means that the relationship between head circumference and maximum long jump is moderately weak.

2 I notice that quite a lot of points are a long way from the line of best fit. This means the relationship between the number of hours playing video games and the number of credits earned in NCEA Level 1 is weak.

3 I notice that all points are close to the line of best fit. This means that the relationship is strong.

4 I notice that most points are fairly close to the line of best fit. This means that for cars, the relationship between the price of a model and the number sold is moderately strong.

5 I notice that quite a lot of points are a long way from the line of best fit. This means that the relationship between new car weight and price is moderately weak.

4 Uneven scatter (p. 39)

1 I notice that below $40,000, most points are close to the line of best fit. Above $40,000, quite a few points are some distance from the line of best fit. This means that below $40,000, the relationship between price and weight of cars is strong, and above $40,000, the relationship is moderately weak.

2 I notice that for cars of under 1500 kg, some points are further from the trend line than those for cars over 1500 kg. This means that below 1500 kg, the relationship between weight and fuel consumption is strong but for cars of more than 1500 kg, the relationship is very strong.

5 Unusual features (pp. 40–43)

1 I notice that one person has a much bigger arm span than would be expected for their height. They are 156 cm tall, but their arm span is 183 cm. I would expect a person who is 156 cm tall to have an arm span of about 156 cm. They are out of proportion compared with the rest of the sample because their point is a long way from the line of best fit.

2 I notice that there is a group of three students who are all very tall (196, 196, and 198 cm) and also very accurate goal shooters (75%, 81% and 80% respectively). Their points are all close to the line of best fit, whereas the points for the rest of the sample are more scattered. The three students who form the group may all be members of basketball teams. Because it is an advantage to be tall in basketball, they are likely to have chosen this sport, and because they are in teams, they practise shooting goals, so they are more accurate than the rest of the sample.

3 I notice that there is one point which is a long way from the line of best fit. One student spent 29 hours per week playing video games, but still managed to earn 133 NCEA credits. This person must love playing video games, but also be very clever. The line of best fit suggests that a person spending this amount of time playing video games should get about 73 credits. The other two students who earned over 120 credits played video games for less than 10 hours per week.

4 I notice that there is one car which costs $80,000, but is only 1600 kg, so this car is unexpectedly light for its price. This point is a long way from the line of best fit. From the line of best fit, I would expect that a car that costs $80,000 would weigh about 2300 kg. Most cars which weight 1600 kg cost only about $40,000.

5 I notice that there is a group of three cars that cost between $8000 and $10,000, even though they are at least 27 years old. Most cars of about that age cost about $1000. For $8000 to $10,000, a buyer would expect a car that was 10 to 14 years old. The three cars in

ISBN: 9780170371636

the group might be very special models that have extra value as vintage cars.

6 There are no unusual points or groups on this graph.

6 Cleaning the data (pp. 44–45)

1 1.65 cm is not a feasible height for a Year 11 student, but 165 cm is, so assume the measurement was made in metres, and change 1.65 to 165 cm.

2 No unreasonable heights for a Year 11 student, so don't remove any values.

3 Remove the 277 cm because it is not a feasible height for a Year 11 student.

4 All the other measurements are rounded to 0 dp except for 151.5, so round that to 152 cm and leave it in the data.

5 -21: You cannot have a negative number of texts sent.
5000: This is an unreasonable number of texts to receive.
2.3: Number of calls made must be a whole number.
17 min and 22 min: The rest of the data is in number of calls made or received, not the time they took.

Practice tasks (pp. 51–62)

Practice task one (pp. 51–54)

Variables

My variables will be the distances that people can hop from each foot. We will use a tape measure to measure the distance in centimetres.

We will ask each subject to place one foot with their toe touching a marked line on the floor, then to hop as far as possible using that foot. We will record the distance between the line and where their toe lands, and which foot was used. We will then ask them to repeat it with the other foot. After they have hopped using each foot, we will ask each person whether their right or left foot is dominant (normally the first foot used when a person starts walking).

Variation

In order to avoid variation we will have the same people giving the instructions, using the stopwatch, and recording the data throughout the data collection.

We will make everybody do this in bare feet, because the type of shoe worn could affect how far they can hop. We will also be able to measure where the end of their toe is more accurately.

We will also take people to a quiet room because having classmates talking and calling out might affect how far people can hop.

We will not ask about which foot is dominant until after we have done the timing, so people don't know why we are doing this. If they did know they might favour one foot or the other.

Sampling

We have been given data from 30 students from Paradise High School. 30 is a suitable number for a sample that should give a clear result. We have no information on how the subjects were selected.

How the data was recorded

Heidi gave the instructions to each person, Pania did the timing, and I wrote down the results in a table with the following headings:

	Time (s)		
Name	Right foot	Left foot	Dominant foot

Data display

A scatter graph was the best way to display this data because it let me show both variables at the same time. It also allows a line of best fit to be drawn which is helpful in discussing the data.

The relationship between distance hopped with dominant and non-domiant feet

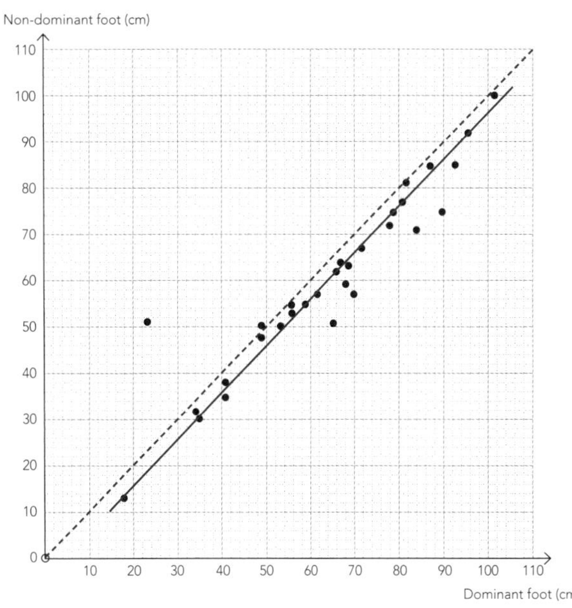

ISBN: 9780170371636

Description of the relationship

There is strong a positive relationship between the distance a Year 11 student can hop on their dominant foot and how far they can hop on their non-dominant foot. This means that as the distance a Year 11 student can hop on their dominant foot increases, the distance they can hop on their non-dominant foot increases also. This does not surprise me, because I would expect that the distances a person can hop would be similar for both feet.

The relationship is strong because most points are close to the line of best fit.

The relationship is linear, because I could rule a straight line of best fit through the data.

I notice that people can usually hop further on their dominant foot than on their non-dominant foot. I have drawn a dotted line on the graph, which is where the points should be if people could stand for the same length of time on each foot.

Only two students hopped further on their non-dominant foot. Chloe hopped 49 cm with her dominant foot and 50 with her non-dominant foot. These values are very close. Jess hopped 23 cm on her dominant foot and 51 cm on her non-dominant foot. Everybody else was able to hop further on their dominant foot.

Unusual features

There are two points which are unusual. The first is from Jess. Using her dominant foot, she could hop less than half the distance that she hopped on her non-dominant foot. This point is unusual because it is a long way from the line of best fit. Perhaps she had injured or recently broken her dominant foot or leg.

The second unusual point is from Fred who could hop only 18 cm with his dominant foot and 13 cm with his non-dominant foot. This is in line with the line of best fit, but the distances are much smaller than for any other student. Perhaps he has a physical disability which made hopping difficult for him.

I did not remove any pieces of data.

Conclusion

I have found that there is a strong relationship between how far a Year 11 student at Paradise High School can hop with their dominant foot compared with their non-dominant foot. Students who can hop a long way on one foot tend to be able to hop for a long way on the other foot. I have also found that students tend to be able to hop further on their dominant foot than their non-dominant foot.

Improvements

- I could have more confidence in my results if I took a bigger sample.
- It would also be interesting to know if this relationship is the same for all ages of students and adults.

Practice task two (pp. 55–58)

Variables

My first variable will be the height of each person. We will measure the height in centimetres.

We will stick a tape measure to the wall, with 0 at floor level. We will ask each person to stand against the wall, to stand tall, but with their feet flat on the ground. We will measure the level of the top of their heads. This will give us a measure of height.

My second variable will be the time each person takes to run 100 m in seconds.

We will take everybody out onto the running track and mark out exactly 100 m. One group member will be at the start of the 100 m, and at the instant they start the runner, they will drop their hand so the timer, who will be at the finish line, can see when to start the stopwatch. They will stop the stopwatch at the instant the runner crosses the line.

Speed is distance/time, and because the distance is the same in every case (100 m), the time taken will give us a measure of speed. Shorter times mean faster speeds.

Variation

In order to avoid variation we will have the same people giving the instructions, measuring height, using the stopwatch, and recording the data throughout the data collection.

We will measure heights with bare feet, because shoes and socks have different thicknesses.

We will measure the top of their head using the right angle formed by a book, to ensure that we get horizontal measurements of the top of the heads.

We will do all the speed measurements on the same day, and in the same direction on the track. This should mean that all subjects will run in the same track and wind conditions.

We will ask all subjects to run in bare feet, so nobody has an advantage from their sports shoes, because these vary.

ISBN: 9780170371636

Sampling

We have data from 30 boys from Paradise High School; 30 is a suitable number for a sample that should give a clear result. We have no information on how the subjects were selected.

How the data was recorded

Penny gave the instructions to each person, Ruby measured the heights and started the 100 m runs. I did the timing, and I wrote down the results in a table with the following headings:

Name	Height (cm)	Time to run 100 m (s)

Data display

A scatter graph was the best way to display this data because it let me show both variables at the same time. It also allows a line of best fit to be drawn which is helpful in discussing the data.

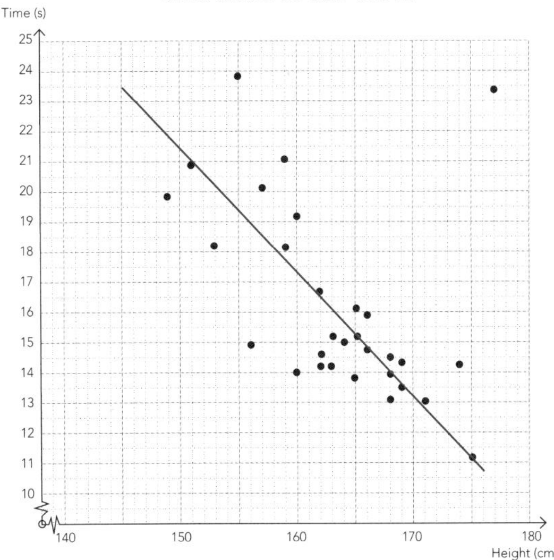

Description of the relationship

There is a fairly strong negative relationship between the height of a Year 11 boy and how long he takes to run 100 m. This means that as the height of a Year 11 boy increases, the time they take to run 100 m decreases. This does not surprise me, because I would expect that taller people have longer legs, and therefore they can run faster.

The relationship is fairly strong because most points are fairly close to the line of best fit. Some are very close but some are some distance away.

The relationship is linear, because I could rule a straight line of best fit through the data.

Unusual features

There is one unusual point. Henry was 177 cm but he took 23.4 seconds to run 100 m — his was the second longest time. I think it is unusual because it is a long way from the line of best fit. Most people who were tall like Henry could run 100 m in less than 15 seconds. Perhaps he was injured, had a disability, or was just very unco-ordinated.

I did not remove any pieces of data.

Conclusion

I have found that there is a fairly strong negative relationship between the height of Year 11 boys and how long they take to run 100 m. This means that as the height of a Year 11 boy increases, the time they take to run 100 m tends to decrease.

Improvements

- I could have more confidence in my results if I took a bigger sample.
- Accurately timing how long it takes to run 100 m is quite difficult.
- It would have been good to have a bigger group so that we could have had two people timing with stopwatches so that we could have averaged their results. This would have given more accurate times.

Practice task three (pp. 59–62)

Variables

My first variable will be mouth size which I will measure in millimetres. I will ask each person to open their mouth as widely as possible, and then use a ruler to measure the vertical gap between their teeth. This will give me a measure of mouth size.

My second variable will be how long it takes for a person to eat a dry Weet-Bix and then drink a glass of milk. I will use a stopwatch to time each person.

Variation

We will get the same person to do all the mouth measuring and the same person will do all the timing. We will also make sure that the measuring is done in the centre of the mouth, because the gap would be smaller towards the sides of the mouth.

For the second variable, we will make sure that everybody has a whole Weet-Bix, and we will not use broken ones. We will also measure out exactly 200 mL of milk for each person, so they all have the same amount. We will also test each person at about the same time of day — in the hour before lunch. It would probably be more difficult to eat a dry Weet-Bix if you were full after lunch. We made sure that people did not start to drink the milk until the Weet-Bix was finished.

Sampling

We have data from 30 students from Paradise High School; 30 is a suitable number for a sample that should give a clear result. We have no information on how the subjects were selected. We assume that none were gluten or lactose intolerant.

How the data was recorded

Amy gave the instructions to the subjects, Jack measure all the mouths, Sam did the timing, and I wrote the data down in a table.

Name	Mouth size (mm)	Time taken (s)

Data display

A scatter graph was the best way to display this data because it let me show both variables at the same time. It also allows a line of best fit to be drawn, which is helpful in discussing the data.

The relationship between mouth size and time taken to eat a dry Weet-Bix

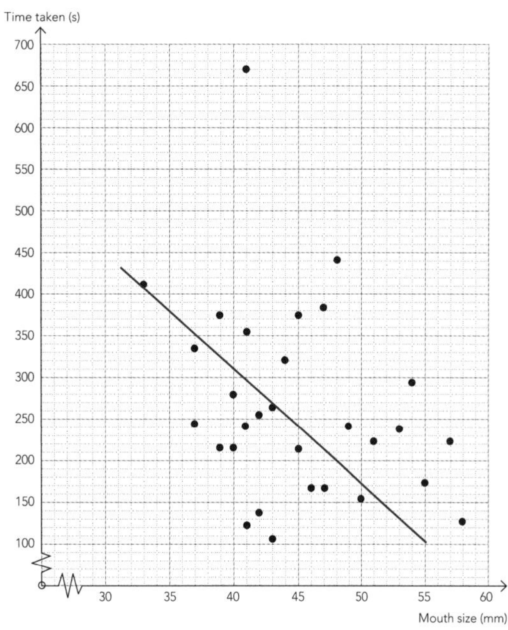

Description of the relationship

There is a weak negative relationship between mouth size and how long it takes to eat a Weet-Bix and drink a glass of milk. This means that as mouth size increases, the time taken to eat a Weet-Bix and drink a glass of milk decreases. I think the relationship is weak because there are lots of points which are a long way from the line of best fit.

The relationship is linear, because I could rule a straight line of best fit through the data.

Unusual features

There is one unusual point: Zoe's mouth was 41 mm, which is about average, but she took 668 seconds to eat the Weet-Bix and drink the milk. Maybe she really doesn't like Weet-Bix or milk.

Conclusion

There is a weak negative relationship between mouth size and how long it takes to eat a Weet-Bix and drink a glass of milk. This means that people with larger mouths took a shorter time, so they tend to be able to eat a Weet-Bix and drink a glass of milk faster than those with smaller mouths. However, this is not a strong relationship, so this tendency is weak, which means that some people with smaller mouths will be able to eat a Weet-Bix and drink a glass of milk quickly, and vice versa.

Improvements

- I am definitely unsure of my conclusion because the relationship was weak. If I needed to be more sure of my conclusion, I would take a bigger sample.
- Measuring the gap between the teeth when the mouth is open may not be the best measure of mouth size.
- If we repeated this, it would be interesting to find out if measuring the width of the mouth gave better results.
- It would also be interesting to know if there is a difference between boys and girls for these variables.

ISBN: 9780170371636